PORTRAITS OF EARTH

PORTRAITS OF EARTH

FREEMAN PATTERSON

KEY PORTER BOOKS

CANADIAN CATALOGUING IN PUBLICATION DATA

Patterson, Freeman, 1937–
      Portraits of earth

ISBN 1-55013-030-7 (bound)
ISBN 1-55013-805-7 (pbk.)

1. Patterson, Freeman, 1937– 2. Landscape photography.
I. Title

TR660.P38   1987   779'.36'0924   C87-093672-7

Key Porter Books Limited
70 The Esplanade
Toronto, Ontario
Canada M5E 1R2

Design: Keith Scott
Editing: Susan Kiil
Editorial assistance: Michael Clugston
Typesetting: Compeer Typographic Services Ltd.

Distributed in the United States by Firefly Books

Printed and bound in Hong Kong

97 98 99 00 6 5 4 3 2 1

Contents

For my editor and friend
Susan Kiil
who is committed to excellence
and to Earth.

The rising sun makes no sound. As day breaks on the Great Sand Sea in Africa's Namib Desert, there begins a symphony of the purest silence I've ever heard. The notes are played out in hues, as the rising sun plays its soft tunes over the early morning dunes. At first the colour is imprecise, delicate. Suddenly a line of crimson appears, a dramatic major motif that lingers in the air. Then, like the string section awakening, the lesser dunes pick up the theme and toss it back and forth, in swirls of deepening colour. It's a moment of total, personal immersion in the world around me – a depth of involvement that produces a sense of oneness with Earth itself.

We don't always have the opportunity to become as familiar with a distant earthscape as I have become with this one. I live on the edge of a forest, near the Saint John River in New Brunswick, Canada. In winter I skim over the ice on my skates, and in summer I paddle my canoe over the same waters. Sky and water seem to merge in reflected images at times, in which I imagine I can see everything from the ice floes of Ellesmere Island, to the breaking surf of the Pacific Coast, to the floral profusion of Namaqualand in southern Africa. In my photographs near home, I am able to travel the world over, in my mind.

The better we know a person, the more likely we are to capture those characteristic expressions that describe the whole person. In the same way, the better I know an earthscape, the more I feel I can convey its character in a photograph. The photographs of Earth that I like best are like the portraits of an old friend – they reveal the essence of Earth itself.

Freeman Patterson

A VIEW OF EARTH

We are, each of us, passive participants in an explosion still in progress. The sparks continue to fly, and in its rhythm, the universe is still expanding from that ancient blast. . . . We are, as they say, star dust, by-products of the Big Bang. In this light, everything is connected, all creation evolving from the same source, in a process which continues. We are kin to the stars, part of a universal family.

CHRISTOPHER NEWBERT

Earth

Earth is our home. It is a small planet in a solar system about eleven light-hours across, the merest speck in the known universe, which spans sixteen billion light-years. Our sun is one of 100 billion stars that make up our galaxy, the Milky Way. Astronomers have detected another 100 billion galaxies beyond ours – for a grand total of ten billion trillion stars and planets. Beyond that, we are still in the dark.

Because light travels at such a high speed, the sun rises eight and a half minutes before we see it at the horizon, but by then it is no longer there. If we make a picture of the sun, we are photographing the past. Every photograph of it is out of date at the moment of exposure. When we observe the night sky with the moon, the planets, and the stars, we are looking, simultaneously, at an incredible number of different moments in time. We see the moon as it was one and a quarter seconds earlier, the planets as they appeared minutes or hours before, a neighbouring star as it was four years ago. And that's only the beginning! If we had a telescope powerful enough to study a solar system in a very distant galaxy, perhaps eight billion light-years away, whatever we could see today might have already become extinct. Conversely, somebody viewing Earth today from a galaxy a mere two million light-years away might observe the activities of our earliest ancestors. Or, given the technology, a photographer in one of the galaxies of Coma Berenice (where light from the Mesozoic era is now arriving) could photograph Earth's dinosaurs alive and well.

In the vastness of the universe, our planet is smaller than a grain of sand in the desert. But we perceive Earth as large and permanent because we are so much smaller than it is, and our lives so brief by comparison. Earth's size and permanence relative to our own makes it possible to plan our futures. Yet, like us, Earth was born, is living now, and one day will die, and between its birth and its death, it will have changed beyond recognition.

The more we learn about the changing Earth, the more we long to escape the usual dimensions of time and space, and to observe our planet as it was and will be. If we could hang a motion-picture camera in space, and have it make a single exposure of our planet once every 5000 years for, let's say, a billion years, the resulting film (just over two hours long) would show Earth inhaling and exhaling, its rib cage rising and falling like our own, prairies lifting, oceans spilling, mountains tumbling, and continents sailing around like toy boats in a breeze. Millions of years ago, the high bluff on which my house now stands belonged to a land mass separate from the continent. Later, it became part of a vast mountain chain as high as the Alps. Today a great river flows by the house through rolling, forested hills. I know all this because of scientific research and technological development; but I also imagine for hours the landscapes of the future. However, all these changes occur so slowly by human time that we can neither see them nor record them. We know about them only because Earth has provided its own record – a geological document that reveals its history.

Earth is the largest living creature in our solar system. The land, the water, the air, and all the things that live on and in them form a gigantic community, an enormous cell. Here fungi, eagles, toads, worms, grasses, mosquitoes, ferns, people, dolphins, spiders, oak trees, and lions – up to ten million separately distinguishable forms of life or species – share Earth's environments. Whatever happens to one part, for good or ill, ultimately affects us all, the whole Earth.

There is constant communication among the billions of Earth's living and non-living parts. We belong to an intricate network of information in which everything contributes and everything responds. We are all related in this vast and complex system, originating together in the moment of creation and freely passing atoms from one to another for billions of years thereafter. Since life began, genetic material has been traded back and forth between plant and animal species, with the new mutations being tested in the laboratories of natural selection – Earth's various environments.

While all of us who live on Earth compete for space and resources like the siblings we are, we also cooperate with each other in myriad ways in order for our cell, our planet, to survive. Through competition and cooperation – putting ourselves first on the one hand and the community first on the other – we achieve balance between order and disorder. Without this balance our cell would die. Early peoples, such as the Bushmen of the Kalahari Desert, the Australian Aborigines, the North American Indians, and indeed most aboriginal peoples of Earth, understood this basic fact and lived in harmony with nature. As keen observers of the natural world, they structured the form or pattern of their lives around the content of their environment. Like all other living things, they *needed* the things that they took from the land, water, and sky, and they took little more.

Yet, history has a beginning and an end. What happens to each of us – we are born, live, and die – eventually happens to planets, solar systems, and galaxies. There is one set of natural laws that applies to everybody and everything. We all share a common fate. Our sun has reached its early middle age. In another six billion years, our dying sun will become a giant red star, consuming Earth. While the end of our solar system is certain, we may have departed this planet long before the cataclysmic event. Already, we have commenced our journey into space – the new wilderness for us, one more vast and wild than any we have explored before. Although we are currently unaware of a planet hospitable to life in any other solar system, and are unprepared to make the journey in any case, six billion years should be sufficient time to search for a new home and to build the vehicles that will take us there.

In the meantime we have Earth, a home of abundant riches and indescribable beauty. Gravity holds Earth's life-sustaining atmosphere in place, the air letting in the sun's energy while protecting us from its harmful rays and solar debris. Because of this protective membrane, we have clouds, rainbows, and sunsets; we have fluctuations in pressure that set particles in motion carrying information in the form of sounds; we have winds that distribute moisture and heat. Aquatic ecosystems support microscopic plant and animal life, as well as the largest of all animals, the whales. In the form of snow and ice, water insulates the land and its inhabitants from winter's cold. An infinite variety of soils cover the continents like patchwork quilts, nourishing plants that, in turn, anchor the soil with their roots and protect it from wind and weather with their canopy of leaves. Both soils and plants provide food and shelter for uncounted species of worms, insects, birds, and mammals that cultivate and fertilize the soil as they dig and burrow.

All of these environments combine with Earth's topography to form earthscapes as diverse as the Rocky Mountains, the steppes of central Asia, the Amazonian rain forests, the Sahara Desert, the polar icecaps, and Pacific atolls. Each of these earthscapes is unique, yet each is connected to all the others, and all are living expressions of Earth's long history of evolutionary development.

Earth's rotation is apparent from the star trails captured during a long time exposure by the camera's open lens. The colours of the trails indicate the age of the stars, and remind us that we are looking not only out into space, but also backward at many different periods in time.

OVERLEAF   Earth's topography is constantly being reshaped by geological forces, water, wind, plant communities, and human activity. These sand banks, built by the flowing water of the Olifants River in southern Africa, can easily be washed away in a flood unless plants anchor the sand with their roots.

In rocky deserts, lichens are the pioneer plants. They slow down the passage of air over rock surfaces, helping wind-borne soil particles and seeds to accumulate in cracks and fissures. After a shower, moisture collects in the cracks and the seeds germinate. If the supply of moisture is sufficient for the plants to complete their life cycle, their leaves and roots will contribute organic material to the crevice, making it a home for additional plants of their own or other species.

The brick-red Precambrian granite hills of Namaqualand in southern Africa are often suffused with tiny red lichens that intensify their hue. Early and late in the day, sunlight reflected into shaded areas from nearby rocks adds to the warmth of colour, often making the setting more colourful than the plants growing there. Chemical staining caused by moisture oozing from rock fissures and the presence of other lichens help to vary the tone and colour scheme.

By creating contrasts of light and dark, sidelighting often helps to convey Earth's three-dimensional forms in the two-dimensional space of a photograph. When natural shapes are isolated from a dark background and/or natural textures are enhanced by lines or dots of blackness throughout, we perceive depth in a flat surface.

Sometimes graphic designs created by light dominate the Earth's own natural designs, or dramatically alter the visual context of natural things. Moving and changing as the Earth rotated, the graphic designs on this rock reached a moment of striking harmony with the sidelighted grass.

This tree has survived harsh desert conditions by living in the shade of a rock wall. Here, protected from strong winds and scorching sun that cause rapid loss of moisture, it sends roots deep into fissures in the rock in search of moisture. The tree's successful struggle to overcome difficult circumstances parallels the life experience of many people, thus making the tree a potential symbol.

In the shade of another cliff the trunk of a much older tree seems almost to have become part of the rock wall. This effect is enhanced by similarities of shape, texture, and line, by the single hue extended through a full range of tones, and by the soft lighting that casts no shadows that would separate the tree visually from the rock.

Both this photograph and the one opposite were made at Augrabies Falls on the Orange River, which rises in the mountains of Lesotho and forms the boundary between Namibia and South Africa before it empties into the Atlantic Ocean. The river carries sand particles eroded from inland rock to the coast where they are deposited in sand bars and banks. Southerly winds may carry some of this sand north and deposit it as part of the spectacular dunes of the Great Sand Sea.

These pictures were made from a great distance and cover large areas. The dark spots on the canyon wall in the opposite photograph are trees, some of them several times my height, which are watered by the spray from the waterfalls. Above, another part of the canyon wall becomes a study in natural design, as water flows over the granite face in a long drop to the gorge below.

The rising sun throws the shadow of a Rocky Mountain peak onto thin clouds drifting between the mountain and my camera. The phenomenon is identical to shadows of trees appearing on a morning mist in a backlighted forest.

Canada's Mackenzie River forms a huge delta before it flows into the Arctic Ocean. Viewed from the air late in the day, the complex pattern is simplified and made more dramatic by backlighting that creates strong contrast between the land and the water.

A single breaking wave becomes a point of emphasis in an ocean, yet enhances the ethereal quality of the entire waterscape. Slight overexposure intensifies the effect of atmospheric haze, softening the shape of a distant vessel that serves as a secondary motif.

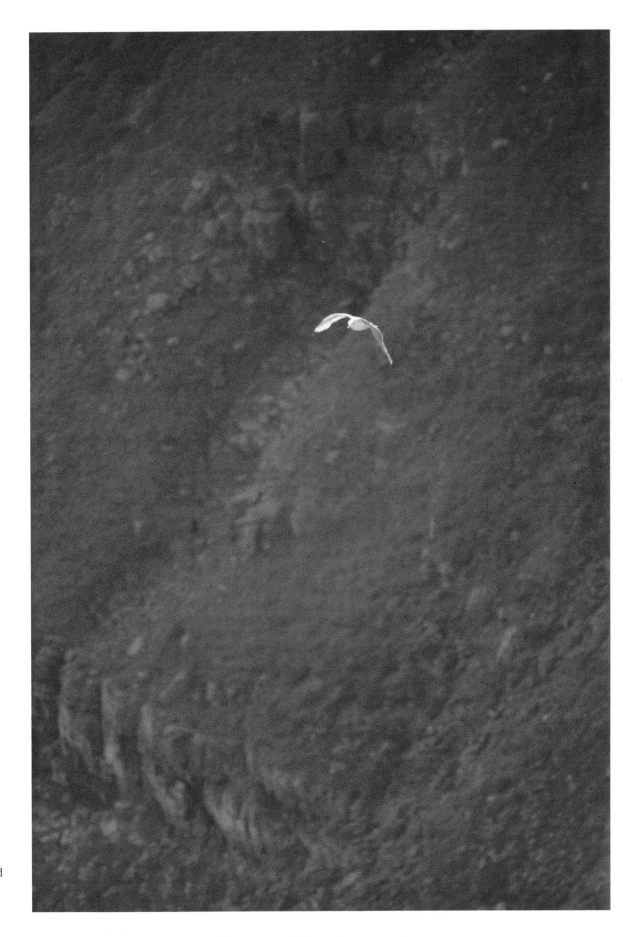

For me, the lone gull soaring against a backdrop of rocky, green cliffs evokes a feeling similar to the wave in the preceding picture. As there was no haze to suggest distance, I put the background out of focus by focusing on the gull and using a shallow depth of field.

When mist or fog obscures some parts of an earthscape, familiar scenes and objects are often transformed into new realities that appeal to our sense of wonder by appearing to defy natural laws – a road winds across the sky or a mountain peak becomes an island of rock suspended in the air.

The natural designs of Earth's topography are a visual history of our planet's development. In this photograph, the rhythmic arrangement of ridges in Alberta's Highwood Pass reveals the ongoing process of mountain building that shapes the west coast of the Americas.

One evening as I photographed cliffs on Grand Manan Island, near the North Atlantic's Bay of Fundy, the rising and falling of sea fog produced three sunrises and three sunsets within the space of half an hour. As the sun disappeared for the last time that day, the fog, flushed with pale pink and magenta hues in the afterglow, provided a delicate contrast to the jagged rocks.

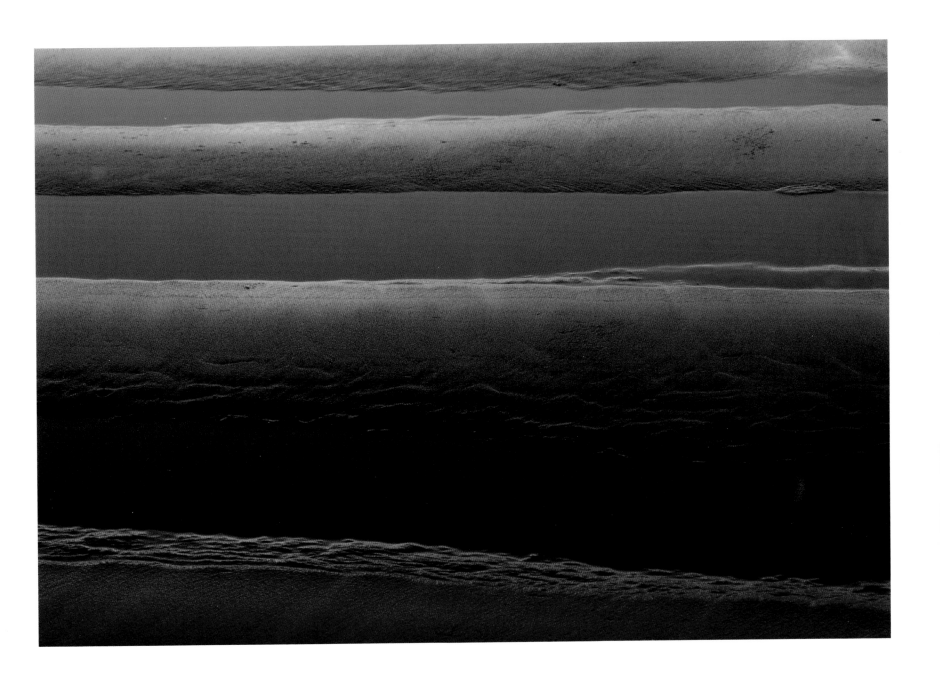

Earth's natural designs may occur in many sizes, suggesting that the forces that create them are unimpeded by scale, as we should be when we view them. This "aerial" photograph of sand bars and channels of water at sunset was actually made from knee level on a Pacific beach.

As pockets of soft rock in a granite slope in Namaqualand (above and opposite) were eroded by swiftly flowing water, small rocks collected in the depressions. There they were swirled around by the water, and like a pestle grinding in a mortar, enlarged the holes and polished their walls.

As the water levels receded over the years, some of these beautifully sculpted holes became collecting pools for rainwater and homes for aquatic plants and frogs. Other holes were partly filled with sand and organic material, and serve as natural pots for grasses and flowering plants.

In the warm light of sunset, Earth seems to smile, transforming even the most rugged aspects of its countenance. Anticipating this fleeting expression, a photographer prepares a composition, and then waits for this brief moment.

A small, muddy creek becomes a great river system when viewed through a telephoto lens in the light and shadows of sunrise. When we free ourselves from our usual perceptions of scale, we can explore the world without leaving home.

Common events in nature sometimes produce uncommon, puzzling designs. Even after we learn how the design was created, a sense of the inexplicable or a hint of mystery may linger on – making us aware that nature has more secrets than we can ever know.

PORTRAITS OF EARTH

The earth does not belong to man; man belongs to the earth. This we know.
All things are connected like the blood which unites one family. . . . Whatever
befalls the earth befalls the sons of the earth. Man did not weave the web of life;
he is merely a strand in it. Whatever he does to the web, he does to himself.

CHIEF SEATTLE

Portraits

Just as we study a face in order to know a person, so we examine an earthscape (a landscape, waterscape, skyscape, or combination of these) in order to understand Earth itself. The countenance of an earthscape changes constantly, expressing the effects of both underlying pressures and external forces. While we may notice a person's eyes, ears, nose, mouth, and other facial features, our first impression is of the whole face – the way the features are put together. So it is with the features of an earthscape; in looking at rocks, trees, hills, rivers, and clouds, we grasp first the total natural configuration, the appearance of the whole. When we meet a person we have not seen for several years, we immediately notice differences. The face we had fixed in our memory is no longer quite the same face, because the person is not quite the same person. So it is with Earth. Its face changes, not only because of the movement of light, but also because Earth itself changes.

However, some of Earth's changes are too quick for our eyes and our memory. We can see a wave approaching the shore, but we cannot remember its exact appearance at any given moment. Even more gradual changes are difficult to apprehend. A few minutes ago I stood by my kitchen window watching the morning sun streaming across the snow. As familiar as this scene is to me and despite my concentration, I can only recall a little of what I saw. There were two birch trees, naked except for the snow on their branches, a greenish-black cedar vaguely pyramidal in shape, a scattering of raspberry canes red in the early light, and large areas of blue shadow where the sun had not yet reached the meadow. These general things I remember. What I cannot recall is the detail – the spaces between the trees, the mesh formed by the twigs of the birches, the varying textures of the snow. Every time I return to the window to refresh my memory the sun has moved, and what I saw in the meadow has changed. The twigs no longer form a mesh, but appear separate and distinct against the snow. The trees themselves seem farther apart, but perhaps I am not observing them from my original spot. The red raspberry canes are going brown. Even as I write this sentence, the change goes quickly on; a cloud is passing across the sun, and a thin, white curtain seems to be descending over the land, muting the hues and tones.

When we observe wilderness, we see physical configurations resulting from the interaction of natural forces. Geological structures, river systems, and plant and animal communities are all interrelated and may reveal a quiet evolutionary development of great duration, or show evidence of a sudden and dramatic upheaval. Even when change has been recent and substantial, we can still sense that the forces that mold Earth are incomprehensibly ancient, and we may perceive something of the infinity of time.

When we turn to domestic landscapes, on the other hand, we are looking to a large extent at the work of our own species, and what confronts us is not so much the character of Earth as our own character; we see ourselves. Because our species is young, and because we have had the technology to alter Earth's surface for such a short period, the patterns of our physical activity are very recent. However, we are clearing, cultivating, developing, and building at such a pace that much of Earth's natural appearance has been obscured or obliterated – with the result that our sense of Earth's history has been drastically shortened. Separated from wilderness in our daily lives and surrounded by highways, buildings, and a plethora of manufactured goods, we have become an introspective species. While we *know* more about Earth than we ever have, we *feel* for it less than our early ancestors did, and our vision of being part of nature has grown dim.

Because the two kinds of earthscape – wilderness and domestic landscape – express different things, the one revealing Earth itself, the other showing the activities of a single species, and because my focus is on Earth rather than on human beings, I have included only a few pictures of domestic landscapes in this book. In wilderness, we have a better opportunity to see what Earth is like.

When we are making a portrait, we usually wait for an expression or gesture that reveals something about the subject's character. Photographing Earth presents much the same challenge. It is relatively easy to capture a reasonable likeness of an earthscape, but much more difficult to photograph an earthscape in a way that reveals Earth itself – the difference between the surface and the inherent reality. However, even an occasional success is enough to make the endeavour a deeply satisfying experience for the photographer who cares about Earth, and caring deeply is crucial to success.

In our day-to-day contact with natural things, we establish emotional bonds. It is as natural to love Earth as it is to love one's human family, and to respond more positively or deeply to some members of Earth's family than to others – to love the ocean, for instance, but to feel uneasy about a dense forest. Different features of Earth satisfy different deeply-felt human needs, or arouse powerful feelings for which there is no easy explanation. The person who cares about Earth is likely to observe it thoughtfully, to see patterns of relationships between things that a casual observer does not, and also to be cognizant of his or her own emotional responses to different aspects of Earth's content. While caring by itself does not ensure clear perception, real awareness of subject matter is impossible without it.

A photographer, like a painter or sculptor, is a craftsperson. But while expertise with the tools and techniques of photography is fundamental to artistic achievement in that medium, it is no guarantee of it. If there is one agent that transforms craft into art, it is a passionate concern, a pervading belief, or a dominating idea that is the heart of, and motivation for, the enterprise. We must never lose sight of this central passion or idea, but we may expand upon it in the process of making a photograph, painting, or sculpture. For a photographer who is both skilled in the craft and painstaking in his or her observations, a strong sense of identification with natural things can be the agent that transforms any kind of earthscape into a compelling portrait of Earth.

A portrait, like a memory, is only a likeness of the original subject; it is a new reality whose truth resides as much in what it evokes as in what it actually portrays. The finest portraits of Earth combine honest documentation with clear statements of the photographer's personal response. They are, in the final analysis, expressive descriptions of a meeting between the subject and the photographer, to which Earth contributes its inherent designs and transitory surface appearance, and the photographer responds by organizing them in the picture space.

However, just as a person seldom communicates successfully by speaking a single word, so a photographer can rarely convey his or her understanding and appreciation of the subject with a single image. A selection of photographs, like a carefully chosen group of words, gives viewers a greater insight into the photographer's response. The photographs in this book show many faces of Earth – its varying topography and environments changing in appearance with light, weather, and the seasons. They are portraits of a very old friend.

Earth and Design

It is when I am at my most observant that I begin to question how accurately I can see or photograph Earth's changing expressions. In one way, my camera sees much better than I do. It records all the details in the scene I have selected – whether I notice them or not. By arresting a particular moment in a particular place, it enables me to observe again and again what that earthscape was like at the split second of exposure, and thus to relive the experience of being there. However, because a camera does not think, it can document only those places and moments I choose for it, in the way I choose. For example, by deliberately overexposing or underexposing, I instruct the camera to bring out certain details in a scene while obliterating others. Even the most realistic images of land, water, and sky have a subjective component; they always represent the photographer's point of view. While a camera, like a pencil or a painter's brush, can be usefully employed as a documentary tool, photography itself is not a science. Like drawing or painting, it is a medium of visual expression. The relevant question for visual artists, then, is not how accurately we can document Earth, but how clearly we can express our personal vision of it.

Natural Design

When we observe the shape of a flower or the lines on its petals, we may see them first as things-in-themselves and remark on their beauty; but as we watch insects travelling over them, we soon come to realize that the shape and lines have a function. They attract the insects to the blossom and guide them to nectar or some other worthwhile prize, and in the process the insect fertilizes the plant by transferring pollen to the female cells. As we wander along an ocean beach we notice that the ripples in the sand seem almost like waves arrested by a high-speed camera. Here the ripples are close together, there they are farther apart; in some places they are very pronounced, in other places nearly imperceptible. Because the patterns are beautiful in themselves, we may stop to photograph them without giving their structure any more thought; but the reasons for the patterns will vary. The sand granules may be larger and heavier on one stretch of the beach than on another, or here and there large rocks in the water may break the impact of the waves before they reach the shore, or the land may not incline to the water everywhere at the same angle. As we begin to observe carefully, the ripples of sand no longer seem isolated, and we realize that the varying rhythms in their patterns and textures reveal the history of the beach.

The intrinsic or *natural designs* of Earth are combinations of shapes, lines, and textures that started to take form nearly five billion years ago, and they have been occurring and evolving ever since. They focus our attention on Earth itself, on its physical content and processes. They tell us what Earth has experienced, they describe what is happening now, and they hint at what is to come. They are a record of historical accidents and evolutionary developments, of mutation and natural selection. By observing natural designs we are able to get behind transitory appearances to the cosmic character of our planet – the connectedness of all living and non-living things.

While all natural things are "ordered," they are not necessarily "orderly" in the human sense of being neat and tidy. There is a randomness about nature; leaves scatter here and there seemingly by accident, or a few seeds germinate from the thousands that a plant produces. Yet, this seeming randomness bespeaks an invisible but pervasive order to which we must be sensitive when we photograph Earth. The movement of air, the shape and weight of the leaves, and the incline of the ground all influence where the leaves will fall. Nothing in nature is purely accidental. But, nothing is entirely preordained either. Perhaps the leaves will fall earlier and be blown farther, if a strong wind comes up as soon as the leaves are ready to fall; but the wind may not blow, or a heavy rain may change the order of things. Photographers who recognize that natural order allows for disorder, that the element of chance is central to the functioning of nature, are likely to be inquisitive about the reasons for natural designs and to document Earth more accurately because of this.

Graphic Design

The natural designs we see and photograph are made visible by light. Although light sometimes seems to create designs of its own, it is actually difficult to find any pure examples of such designs on Earth. Because these inherent designs are present everywhere – in land, water, and air – light is always reflecting off or influenced by something that already has form. However, when light has a dramatic or pronounced effect on the appearance of an object or earthscape, we may find it useful to speak of the design created by light or, more appropriately, of *graphic design*, which acknowledges the strong effect light can have on the appearance of inherent designs.

Because light never stands still, it is constantly altering the appearance of natural objects and situations, which is to say that it affects the impressions we have of their shapes, lines, textures, and perspective (the appearance of distance or depth), and the way these design elements have been put together. Within moments, a flat expanse of snow may be transformed by a shaft of light into a shimmering, diamond-studded mosaic. A passing cloud may mute the electric blue of a northern fiord or a prairie slough to a delicate grey. Light may leap and dance across a range of mountain peaks, then ricochet off lakes in the valleys below; or its movement may be so imperceptible that hours will pass before we realize, in a sudden moment of awareness, that nothing looks the same any more.

Sunlight on a clear day produces quite different effects from sunlight filtered through clouds, which deflect and disperse light rays. On a sunny day a rocky hillside may be distinguished by brilliant highlights and deep shadows that enhance the natural structure of the rocks, creating powerful shapes. On a cloudy day with the light more evenly distributed, strong contrasts will be missing, and from even a short distance away the shapes of the rocks may actually seem to merge with those of surrounding stones or grasses.

The direction in which the light shines strongly influences visual contrast, and can substantially alter the appearance of the natural designs of an earthscape. A large boulder illuminated from the front may seem flattened, its shape and texture somewhat lost against the background. When the sun moves, or we move, so that the boulder is illuminated directly from one side or the other, the edge of the side nearest the sun is sharply delineated. Now the rock appears to be a large natural sculpture, having texture and depth. If we walk behind the boulder, so we see it silhouetted against the sky, it will again seem textureless and flat, but its shape will stand out strongly, enhanced by the dramatic contrast between dark and light. In all these ways the direction of light alters contrast and changes the appearance of the rock by visually enhancing or diminishing aspects of its natural design. If, however, we visit the same boulder on an overcast day, when all sides are illuminated evenly, the appearance we see from any position is more nearly that created by geological forces, climate, and weather, without the graphic alterations of light and shadow.

Whether the sky is clear, partly cloudy, or overcast, there will be different areas of brightness in every earthscape. Often these differences are caused by the ability of Earth's various surfaces to reflect or absorb the light rays falling on them. A piece of quartz stands out in a field of rich loam for precisely this reason. At other times, a field that is by nature more or less the same everywhere will appear darker in one area and lighter in another because of a passing cloud or a nearby hill, or because of slight differences in the angle of its surface. An intensely dramatic variation occurs when light bursts through dark clouds, briefly illuminating part of a lake or an ocean. However they are caused, these contrasts of brightness are at the core of visual perception, creating visible shapes, lines, textures, and perspective. In fact, without visual contrasts, it seems almost certain that animals would never have evolved eyes, or would have abandoned them long ago in favour of developing more useful sensory organs, perhaps enormous noses or grossly enlarged ears instead.

People and birds (whose noses are small or missing altogether) are able to distinguish colour contrasts far better than other animal species. Like contrasts of tone, colour contrasts provide information about land, water, and sky, and produce shapes, lines, textures, and perspectives that influence our response to earthscapes. For example, colours help us tell time. Because of natural pigments in plants, we know spring by a bright green field, summer by darker greens, while a brown field tells us that the growing season is over. Similarly, changes in the colour of light communicate the time of day in both our lives and our photographs. We know that "warm" hues (yellow, orange, and red) dominate when the sun is near the horizon, and that "cool" or bluish light marks the middle of the day. These changes in the colour of light also influence our emotional response. A blue-grey mountain peak that fails to attract more than a brief glance at noontime may have us racing for our cameras a few hours later when the rock glows magenta-pink in the last rays of the setting sun. Generally, strong primary colours quickly attract our attention, but the less emphatic secondary hues may hold our interest longer once we become aware of them.

The mixing and desaturating of hues combined with the lightening and darkening of tones occurs naturally and continuously as the hues and tones of light meet Earth's myriad natural pigments. For example, the natural triangular shape of a mountain peak does not change from midday to sunset, but it may appear to change because of shadows created by the moving position of the sun; and this new shape may be a different colour than the previous one because the colour of the sun's rays has changed as well. Both the new shape and the new colour may affect our response to the mountain scene.

In addition to causing visible changes in an earthscape, different light conditions make us *feel* differently; they alter our moods. How we feel may well influence how we perceive things, including physical objects and situations. Thus, because light affects both the perceiver and the things perceived, it has a critical bearing on the photographs we make of Earth. Good earthscape photographers spend a lot of time waiting for the lighting they want; their patience is almost as important as their technique!

Personal Design

The graphic designs created by light, like Earth's natural designs, are the raw materials, as it were, for the pictures we make. However, there is a third kind of design – the design we create in the camera viewfinder – which we can call *personal design*, picture organization, or composition. It is through this kind of design that we express our personal response to Earth.

Because photographers compose with objects that have an existence of their own (unlike painters, who begin with an empty canvas and build or construct on it), there is a necessary and direct relationship between Earth's natural and graphic designs and the designs we create with a camera. Painters may decide to eliminate the lines on the petals of a flower, or to create a flower like none that has ever existed, or to juxtapose species that do not grow in the same habitat; but photographers, however much we may alter the normal appearance of things by our choice of camera position, lenses, filters, and so on, have a working connection with physical reality and natural order. We must make use of what we have been given, and how we work with it will demonstrate the degree of our understanding and response to it.

Since we perceive the total configuration of a scene or an object before we notice its parts, we tend to overlook the details of a situation that fails to interest us. However, if the scene moves us or attracts us enough that we want to photograph it, we begin to analyse its parts and the way they are put together. When I decide to photograph an earthscape, perhaps a field of brown grass sweeping away to rippling grey water and a leaden sky, I begin my examination of its forms by peeling away the labels "grass," "water," and "sky," and observe them as shapes, lines, textures, and perspectives. This helps me to understand why subject matter I have observed hundreds of times before can take on a new meaning for me in a purely visual sense; then I'm able to recognize how the subject matter is expressing itself.

As I study the scene, I may decide that it is the overall texture of the field (the rough but intricate weave of millions of lines of grasses) that attracts me, or the shape of the field, rather than the texture of the grasses. Or, perhaps all the grasses appear as a multitude of delicate oblique lines from where I stand, giving a subtle, but pervasive dynamic to the scene. Or, I may realize that texture, shape, and line are all less important than I originally suspected, and that the brown grasses evoke old associations and memories of days long ago on my parents' farm when I drove our dairy herd down wandering paths and across old fields to the pasture on the hill. In any case, I have to recognize the visual forms so I can select and emphasize the ones that will help me to create or recreate my response to the original scene. If I find the texture of the grasses compelling, I will probably choose to concentrate on that and to eliminate the shapes, lines, and textures of the water and the sky from my composition. But, if the field evokes strong memories of my days on the farm, I will leave them in – because the sense of distance was always with me when I made the twice-daily trip to the pasture. So, I make visual choices, and with these choices I reveal not only something of Earth, but something of myself as well.

Once I have selected the visual elements I want, I have to organize or arrange them in the camera viewfinder by deciding on the proportion of space for each, how to balance them effectively (by their position and orientation within the frame), whether or not to make one element dominate the others, and so on. In doing this, I am confronted by the shape of the picture space, which is rectangular or square, and by the edges of that shape. A viewfinder (and a photograph) does not allow for peripheral vision; space is rigidly truncated. As I compose the picture, the edges of the viewfinder force me to create shapes that do not exist in the scene, causing me to consider how they alter the natural and graphic designs of the subject matter. Perhaps I will need to zoom the lens or move the camera nearer to important material in order to maintain the significance I want it to have (so the shape of a tree will be larger in the picture than surrounding shapes and the texture of twigs and branches will be more apparent). Or, perhaps I will want to move farther away (so the shapes of individual flowers will be reduced in size and more picture space will be devoted to the vast carpet of bloom, as on pages 150 and 151). Or, I may decide to move the camera or aim the lens either higher or lower, and to one side or the other (so that I can include the dynamic ridge line of a dune and suggest the power of wind, as on page 69).

Moving the camera position is often the easiest way to influence the appearance of shapes, lines, and textures, but by itself this method may be insufficient to convey a sense of depth, that is to show the distance or perspective of an earthscape in the two-dimensional space of a photograph. Generally, we understand or "see" distance because of the deformation of space, or distortion. If I am photographing a deer and it runs away from me, it will soon appear much smaller than it did moments before. I do not assume that the deer is shrinking, but rather that it is increasing the distance between us. While this is initially a learned or acquired perception, once we understand it we can see in no other way; in fact, we transfer our normal perception of depth or distance to a photograph or painting automatically, provided the picture includes objects whose comparative size we know (a tree compared to a mountain, for instance). The size of these objects relative to each other in a picture makes us perceive distance where none actually exists. This is very convenient for photographers, of course, because often much of the work in conveying distance is done for us; but when we want to alter the relative size of objects even more, a change of lens as well as a change of camera position may be required.

If we use the lens we already have on the camera and move a little closer to important foreground objects, perhaps some flowers, we can enlarge them enormously without noticeably increasing the size of the background – perhaps some hills – thus changing their relative sizes in the picture space and increasing the impression of distance between them. However, by using a shorter-focal-length lens we can come closer to the foreground flowers (enlarging them while keeping them in focus). The resulting distortion of the relative size of the flowers and the hills will cause viewers to see great distance. We can make the effect even more pronounced if we switch from the horizontal to the vertical camera position, because the vertical position allows us to move in more closely to the foreground objects, thus deforming space even more. Often, positioning the lens not only close to, but also slightly above the foreground, helps to separate the foreground objects from the background, thus showing viewers enough space between the two areas to let them "feel" the distance (see the photograph on page 73). If we want to decrease the perspective of an earthscape, we can do the opposite. That is, use a longer-focal-length lens and move back from or eliminate important foreground material.

There are other factors that affect our perception of distance as well. We use most of these in our daily lives and can employ them effectively in picture organization to increase or decrease perspective. For example, clearly defined or outlined objects will usually seem closer than objects that – because of focus, depth of field, haze, or other natural factors – are not as sharp. In a range of misty hills the nearest hill looks nearest because it is most sharply defined. However, if weather conditions change and the nearest hill becomes more obscured than the others, some of the sense of perspective will be lost – even though we know the hill is nearer than the others because it is lower in our field of vision. In fact, most objects located in the lower part of a picture tend to appear closer than those in the upper part. This too corresponds with real life, where near objects are usually lower in space than far ones. The presence or absence of a horizon also affects our perception of distance, as does the location of the horizon in the picture space. Generally, if the horizon is placed high in the space it will seem farther away than if it is placed low, but a low placement will evoke a greater feeling of height or space, because our attention is focused on the sky. Also, the overlapping of one object or part of a scene by another suggests depth, even when a small object blocks part of our view of a larger one. These and other clues to perspective from daily experience can be utilized in picture organization to increase or decrease the sense of distance, and most of them can be achieved by changing camera position, by changing the lens, or by doing both.

Camera position and lens choice are just as important when we want to arrange shapes, lines, and textures, because the choices we make can substantially affect how these visual elements of an earthscape appear in a photograph. For example, when we replace a wide-angle lens with one of longer focal length and eliminate foreground objects from a composition in order to reduce perspective, we increase the expressive power of shapes (and often texture, as well). Because there are now fewer areas or objects in the picture space, the shape of each is larger and more important; and the reduction or elimination of perspective will help simplify the composition (see the photograph on page 80). If the impact of one or more of the shapes needs to be increased or decreased, or the shapes need to be placed in better balance, we will have to make further adjustments to the camera position, choose another lens, or both.

Another factor that can decisively affect composition is usually best considered after all the other design elements have been arranged in the picture space. This is depth of field – the span of distance in front of and behind the point of focus that appears fully sharp. Because objects that are in focus usually have more visual importance than those that are not, we can increase or decrease the power of shapes, lines, textures, and perspective by varying the depth of field. We can have everything in focus by using a small lens opening with a wide-angle lens, or we can zero in on a critical element of the composition and render everything else out of focus by using a wide aperture with a telephoto lens; or we can try for a depth of field between the two extremes, if we feel this is the most visually satisfactory approach. At this point we have to examine the subject matter for visual cues to the best setting and ask such questions as, "Can I show the sweep of the land more effectively by employing full depth of field, or by focusing on a single feature and implying distance by throwing everything else out of focus?" Or, "If I use shallow depth of field to emphasize only the clump of dry grass, will the resulting loss of texture in the snow diminish the feeling of winter?"

Since the amount of light reaching the film will strongly affect the appearance of our subject matter and the visual effect of the entire composition, we must think of the exposure we use not only as a necessary tool, but also as one of the most creative techniques available to a photographer. The most appropriate or effective exposure depends on what we want to accomplish. Often our aim is to show the subject matter as it appears or would appear in "normal" daylight, that is, to expose a scene in a way that clearly reveals its natural and graphic designs. Sometimes, however, we want to retain or create a visual effect or a mood that requires overall lightness or darkness, and, at other times, to emphasize some aspect of a scene by exposing it clearly while allowing other parts to appear very dark or light by comparison. Once we have obtained the meter's recommended exposure for very dark storm clouds, for example, we will underexpose perhaps two shutter speeds or lens openings to keep the clouds dark in the picture, while retaining some detail in them. On the other hand, we will overexpose snow scenes if we want the snow to look white (see pages 126 and 127).

When we are photographing earthscapes, we can control the colour-tone combinations in our pictures through selection of the subject matter and technique. For instance, we can add, change, or eliminate combinations of colour and tone by altering our camera position, by varying the time of day, by lengthening or shortening the exposure time, and by using filters. We can diffuse, soften, or mix combinations by throwing them out of focus and/or using a shallow depth of field, and we can alter their visual importance by increasing or decreasing their size in the total picture area. The important thing, of course, is to have good reasons for the decisions we make.

If a scene is dominated by vibrant reds, oranges, and yellows of middle tone, let's say a maple forest in October, we may decide to make a composition that includes a dark green spruce or pine tree, because its contrasting hue and tone make the warm hues seem even more vibrant by comparison, and we may add a polarizing filter to eliminate atmospheric haze that is lightening the overall tone, softening the colours, and reducing their impact. If we are attracted by the pattern created by a patch of delicate yellow flowers, we may carefully avoid including in our picture any dark spaces between the blossoms that become shapes competing with the overall pattern; then, to express the delicacy, overexpose slightly to lighten the yellows.

In every case our response to an earthscape will be affected by the designs of Earth's natural features with their inherent tones and hues, as well as by the graphic designs produced by the colours and tones of light. So careful observation of both will help us to emphasize the essential visual elements.

Thus, in photographing an earthscape we follow a sequence of steps, a process. First, we observe the scene's natural and graphic designs; then we select and arrange the designs we want in the viewfinder by employing a specific lens and camera position. Next, we determine the depth of field by choosing a lens opening; and finally, we adjust the shutter speed and expose the film to light. While we may vary the sequence, the basic connectedness of the steps is fixed. Just like all the living and non-living parts of Earth, the steps in making a photograph should work together to make an effectively-functioning whole.

Desert Portraits

There is an air of timelessness and infinite space about deserts. Some of them are still undisturbed by roads, buildings, transmission towers, and other constructions of our species. Here, free from these human artifacts, I feel more like a voyager in outer space than I do anywhere else on Earth, and nowhere is my feeling of exploration and discovery more intense than in the Namib Desert of southwestern Africa and on Ellesmere Island in northern Canada.

The Namib Desert

The Namib is one of Earth's most unusual, varied, and starkly beautiful landscapes. This desert stretches south along the Atlantic coast from southern Angola through Namibia to the Olifants River in South Africa, embracing regions whose names haunt my imagination – the Skeleton Coast, the Naukluft Mountains, Kaokoland, Namaqualand, and the Great Sand Sea. Some of these places I have visited time and again; others I have never seen but long to see.

Sunrise transforms Earth everywhere, but nowhere is the transformation more striking than in the enormous dunes of the Great Sand Sea, and nowhere else have I been so aware that the rising sun makes no sound. At dawn one experiences a symphony in pure silence. A few minutes before the sun clears the escarpment far to the east, the sea of dunes begins to glow pink, as if a giant heater buried in the sand had been switched on. At first the colour is so imprecise, so delicate, that what one feels is the expectation created by the distant sounds of an orchestra tuning up in the depths of an auditorium – before the orchestra platform has been elevated to the level of the main stage. Suddenly a line of crimson edges the ridge of the highest dune, a dramatic major motif that lingers briefly, alone in the air. Moments later the motif is picked up and repeated in myriad variations by all the lesser dunes, much as the strings in an orchestra echo, yet alter, the theme first stated by the lead violin. Then, seemingly from everywhere, come the equivalents of woodwinds and brass – great purple swirls, rippling golden textures, and long shadows as black as midnight that sweep through the visual symphony separating and defining its parts while ensuring its harmonic structure. I see all of this and hear nothing, but in my memory the desert sunrise is full of music.

It is not easy to photograph a symphony. Often when I am listening to music I let the whole sound roll over me and through me, but other times I concentrate on a single phrase or instrument, as if I were examining one blossom in a field awash with flowers. This latter approach is useful in the Great Sand Sea, because so much that is visually exciting is happening all at once. While I can experience the overall effect of sunrise in the dunes, I cannot capture the entire impression in a single image – because pictures have edges. So, I usually select one spot – the base of a dune, perhaps, or a high ridge of sand – and spend an hour or more photographing in that location before moving on. By observing and photographing a few parts of the desert well, like picking out and playing the central themes of a musical score, I hope to suggest the character and emotional impact of the whole.

Most mornings when I camp in the Great Sand Sea I roll out of my sleeping bag long before the last stars disappear into the orange glow above the horizon, and grope for a thermos of hot water to make a cup of tea – several cups of tea, in fact. Since the day will be spent in the sun, and heat, and drying winds, I must guard against dehydration. My lightweight backpack is already stuffed with small containers of fruit juice, as well as with raisins, nuts, a couple of hardboiled eggs – and lots of film, of course. I put on loose-fitting cotton trousers, cotton socks, and hiking boots that come part way to my knees. Over my sweatshirt (necessary for the chilly morning temperature) goes a thin denim vest that I made by ripping the sleeves off an old shirt. When the day becomes warm, I will stuff the sweatshirt into my pack, lather my arms with sun-blocking cream, and wear only the vest. I check my tripod and two cameras (one with a 28-85mm zoom lens on it, the other with a 70-210mm zoom; both lenses fitted with a warming filter and a polarizing filter to counteract ultraviolet and spectral light). I put the camera with the longer lens on the tripod and the other one around my neck. Finally, I slip a lens extender (for very long telephoto shots) into an outside pocket of my backpack, double-check to see if I have forgotten anything important, don an old straw stetson, and am on my way. It is half an hour till sunrise.

Walking as swiftly as I can in the soft sand and pre-dawn light, I head for the base of a huge dune about twenty minutes away, though any other direction would be just as good. Before I am halfway there, the sky behind me turns the colour of blood and the sea of sand brightens to an iridescent pink. Immediately, I am in a dilemma. Do I stop now and take advantage of the spectacular early light, or do I press on to be certain of seeing the first rays of sunlight striking the big dune? I decide to try for both. Quickly positioning my tripod and camera, I aim my lens at the sihouette of an ancient camel thorn tree, one of many still alive in the dry river bed. The black lines of the three main branches command my attention, so I adjust the camera position slightly and zoom the lens from 70mm to longer focal lengths until one branch touches the lower left edge of the frame, another the upper right, and the third curves its way to the top of the picture space touching the upper edge to the left of centre. Very near the bottom,

I include the undulating horizon. Satisfied that I have established a strong sense of the tree's natural design by this balancing of lines, I quickly choose medium depth of field (which is sufficient to render both the tree and the horizon in focus), underexpose a little for extra colour saturation, and press the shutter release. Resuming my hike I am soon stopped again, this time by the glowing texture and shape of another massive camel thorn (see page 64), and make several identical exposures of my composition – better to have enough original slides for a variety of uses than to make duplicates later. Then, I grab my tripod and race for the dune. I barely make it in time. No sooner have I climbed a low, sandy knoll with a commanding view of the great dune's soaring, twisting ridge than the sun peers over the distant escarpment. The time for ''tuning up'' is over; the symphony begins.

The first ray of crimson light is like a single note played on the towering crest of the dune. (I use the lens extender behind my long zoom for a 400mm photograph.) Then the note is repeated on the peak of a second dune. (I remove the extender and zoom back to a shorter focal length to include both peaks.) Gradually, the colour expands along and down the ridge lines of the low dunes. (I switch to my second camera with the short-focal-length zoom lens.) Suddenly, the ''music'' erupts in a hundred places at once. (Anticipating this, I had already zoomed back to 28mm and composed the scene in the viewfinder.) Finally, the colour cascades over the whole sand sea in a spectacular crescendo of light. In this vast ocean of silence, there is music everywhere.

Laying my equipment aside, I crouch in the sand – observing the spectacle of morning in an unfamiliar world. Often when I am deeply involved in making pictures, I pause to experience the situation simply and directly. For a few moments I want nothing to intrude on my experience of being there, nothing to interrupt my feelings. When I pick up my tripod and cameras again, I have no sense of regret about the photographic opportunities I may have missed, because I have a better understanding of these vibrant hues and strongly contrasting tones. I have come to realize that all the familiar labels we apply to landscape objects – dunes, hills – have become momentarily meaningless, because here the striking designs created by light often visually overwhelm the natural forms and content, sometimes producing powerful scenes that we might associate with science fiction.

Except for midday in the Great Sand Sea, when the brilliant overhead sunlight casts no shadows and renders the dunes a delicate pink or pale beige (often very beautiful in its own way), the oblique direction of the light creates strong shapes and lines, and enhances texture. To convey the highly graphic appearance of a particular scene, photographers must pay extraordinarily careful attention to composition. Often that means waiting until the sun has changed position and altered shapes or lines in a way that maximizes their impact. To make the picture on page 68, I made an arduous climb through soft sand to the top of a high dune, then waited as the ridges of another dune began to form long parallel rectangles in the slanting light. The lowest rectangle was partly black, the one above it a textured gold, the one above that entirely black. Towering over them all, the curving peak of the dune caught the golden sunlight that illuminated the second rectangle, and above everything was the rich blue sky. In this situation I could see that the natural structure of the land would be more striking if *all* of the lowest rectangle were to become black while everything else in the scene remained as it was. Studying the contours of the surrounding dunes and the position of the sun, I decided that the chances of this happening were good. So, I set up my tripod and camera, placed the shapes in the picture space with meticulous attention to both their proportion and position, determined the most appropriate exposure, and then waited for the final piece of design to fall into place – for the earthscape to attain its moment of greatest visual strength.

There are times and places in the desert, however, when the nature story revealed in the natural forms is more arresting than the graphic designs. One day some friends and I hiked across a stretch of low dunes that terminated abruptly at an almost vertical barrier of sand on which the sun had been beating for hours. The climb was short, but difficult, rather like scaling the red-hot element of an electric stove. After several minutes of extreme discomfort, we cleared the top of the ridge. An eerie panorama stretched out below us. Huge sand dunes towered over a broad expanse of gleaming tiles of dried white mud, studded with the black, twisted skeletons of ancient camel thorn trees. Centuries ago this arid pan had been part of a river that flowed occasionally to the sea, providing sufficient periodic moisture for plants to grow in the river bed. Then enormous dunes advanced across the river, depriving the area of moisture and eventually covering the trees completely. The camel thorns had been baked in an oven of sand for hundreds, even thousands, of years before the dunes moved on, uncovering the blackened and petrified skeletons.

One morning I returned to experience this haunting landscape alone. After I made the photographs on page 70 and 71 from the high viewpoint of the ridge, I descended into the pan. The silence was so complete that I became aware of every rustle of my clothes, and decided to sit under one of the dead trees to stop the sound of my own movements. For the first time in my life I was in a total auditory void. The absence of sound and any human imprints or artifacts produced an overwhelming sensation of timelessness. These days, to be able to be alone in such a wilderness is a rare gift indeed.

Eventually I resumed my original intent – to capture something of a landscape unlike any other I had ever known. I wandered from tree to tree across the flat, white bottom of the dead lake in the blinding noonday light, pausing here to study the mosaic of mud tiles, there to isolate a tree against a distant mountain of sand, and somewhere else to photograph the contorted shadows of the skeletal branches. As I moved away from the trees, I became aware of the surrealistic effect produced by brilliant overhead light – a virtual absence of shadows, even my own. The laws of nature suddenly seemed out of whack. Only an hour earlier the absence of sound had disoriented my sense of time, but now I had the impression of being outside of time altogether.

After several hours I climbed back up the ridge of sand and paused for my final photographs of this alien world. Turning at last to descend the other side, I was confronted by yet another illusion. Far in the distance a lone gemsbok was walking in the silvery haze that hung above a broad expanse of blue water – a mirage. I sat in the sand and watched until the antelope had disappeared into the mist. This vision of an antelope walking above the horizon made me aware of one of our most common illusions – that land and sky, or water and sky, meet only at the horizon. In fact, of course, air meets land beneath our feet. We are always standing on someone else's horizon. I then looked down to see whose horizon I was imposing on, and immediately realized that the patterns of sand ripples on the ridge were only a smaller version of the entire desert structure. This thought quickly transported me to a new viewpoint, and I used my short zoom lens to make ''aerial'' photographs (from my normal height) of the land around my feet. The curving parallel ripples of sand, resembling dunes, told of wind erosion, and provided basic order for my compositions. The criss-crossing trails of insects and lizards revealed other sorts of activity, and reminded me of human paths and roads I have seen when flying over deserts.

Next, I lay on the sand and looked along and across the ripples. Again, they took on the appearance of dunes, but now my low position made the nearer ripples seem larger than the farther ones, and the perspective was enhanced by the presence of a horizon. At any height a horizon lures us, not only beyond familiar physical surroundings to more distant and less familiar ones, but also beyond apparent physical reality into realms of imagination. Early explorers felt the tug of the horizon, and for us it is equally a subject of wonder – the place where ''the beyond'' commences.

In a picture, a horizon forms an edge or line that divides space into an upper and a lower shape, which is an unnatural representation of the world. However, when our eyes roam over a scene, the horizon acts as a point of transition rather than as a sharp dividing line between sky and land, or sky and water. Furthermore, the picture has edges, but the actual scene doesn't. The edges terminate the horizon abruptly, at both sides of the picture. Any line that is stopped abruptly becomes important because it has clearly reached its destination, and therefore has delivered its message. Sometimes a strong horizon is pictorially desirable, but if the horizon in a picture confines us rather than releases us, if it inhibits our sense of wonder, we may decide to eliminate it or to alter its expressive force by repositioning it in the picture space, by throwing it out of focus, or by using some other technique.

Because the horizon is as much a point of psychological departure as a physical one, I always consider carefully where to locate it in the picture space – especially when I want to strengthen or diminish its importance in the overall composition. The most appropriate placement will depend on one or more of several factors, such as the shape and position of the picture format (rectangular and vertical, rectangular and horizontal, or square), whether the horizon is straight, curved, undulating, or jagged, the colour and brightness of the land or water, the colour and brightness of the sky, the presence or absence of clouds and of objects at the horizon. All of these physical considerations, and many others, may have important expressive effects and will influence the final decision.

The horizons of the Great Sand Sea are much more than a matter of camera position, however; they are also the natural expressions of its topography. Here one experiences every conceivable surface orientation or position. There are nearly *vertical* dune faces, flat or *horizontal* surfaces, and countless *oblique* lines and shapes that give much of the dune landscape a strongly dynamic appearance. The sheer variety provides an unparalleled opportunity to study, in a relatively limited area, the visual effects of surface orientation as well as some of the ways it aids in creating microclimates and in establishing various kinds of habitats within an overall environment. These visual and biological lessons can be usefully applied where soil and climate are very different – in mountainous areas, for example, or on prairies.

There is a close relationship between the orientation of a dune surface and its texture. Very steep or vertical sand faces are relatively smooth, whereas horizontal and oblique surfaces exhibit a range of ripple patterns of varying depth. Texture usually refers to the roughness of surfaces, such as sand ripples on a dune or baked mud tiles (see page 174). However, it can also be thought of as the interaction or weave of natural objects on Earth's surface, perhaps a carpet of flowers after a desert rain. Texture is often the dominant visual element in natural situations – stronger than shape, line, and perspective. In fact, in numerous instances texture is so powerful that to look for a shape or line to serve as a centre of interest for a photograph is to destroy the visual impression by diminishing the real centre of interest – the texture.

The roughness of any texture is enhanced by increasing contrast – for example, side or backlighting highlights sand ripples or mud tiles by seeming to darken the spaces between them. However, the appearance of texture is also a matter of distance; it can be created, altered, or eliminated by varying the camera position. Viewed from great height, the rhythmic patterns of sand dunes seem like the threads of a carpet; but close up, these lines become relatively far apart, and cease to be threads. In their place, ripples of sand establish a new, but often similar pattern. However, when examined closely, the ripples disappear and the texture formed by granules of sand comes into focus. All natural textures suggest interaction, cooperation, community, habitat, and environment. Because textures reveal an intricacy and richness of relationships that no other visual element can convey so satisfactorily, they help us to portray nature whole.

As the sun nears the western horizon, the Namib's entire Great Sand Sea is reversed graphically from its morning appearance. Aspects of dunes that were glowing red shapes at sunrise are now replicated in deep grey or black; dark morning shapes appear again in bright orange and crimson. It's as if the visual symphony were being played by a new conductor – one who has decided on a whole new emphasis. As the sun disappears, the dunes again give off a warm, pink glow as they did in the pre-dawn light, but soon it is dark. Night settles quickly over the base camp at the edge of the sand sea, a black so solid and complete that no hint of the desert landscape is visible anywhere; only the barking of the gecko lizards convinces me that the desert is still there. Above me, uncounted billions of stars prick the darkness, some appearing so brilliant and close that I feel I can reach up and take them in my hand. How incomprehensible it seems that all the grains of sand in the desert do not equal in number these grains of light in the sea of blackness.

Ellesmere Island

The Arctic tundra is a cold desert that circles the top of the globe between the polar ice cap and the tree line. In many respects, it is as unlike the Namib Desert as it is possible to conceive, while in other regards it is so similar that I sometimes wonder if the two weren't once good friends who drifted apart over the years.

Tundra means "a treeless plain," even though the Arctic tundra is sometimes rolling or hilly, even mountainous in places, as on Ellesmere Island in the high Arctic of northern Canada. Ellesmere's coastline is probed by long, narrow fiords whose mountain walls are capped by vast fields of ice that extend great distances inland. As in the Namib, the absence of tall vegetation makes the sweep of the land visible everywhere – its vistas as spectacular as they are enormous. There is almost no sense of scale, and the light plays tricks on one's perception. Nearly everything is either very bright, like ice and clouds, or very dark, like rocks. There are few middle tones, and the remarkably clear air gives everything a hard edge. There is none of the softening of distant objects that aids in judging the span between one boulder or ridge and the next. Because distances are so hard to judge, the mind feels free to wander over every line of a great cliff face, to imagine scaling the tongue of a glacier that hangs out of a valley high above a fiord, or to set sail on a huge blue cake of multiyear ice. Everything is a lure to exploration, and every imagined hike or climb seems as possible as it might to a child caught up in a book of adventure stories.

In an Ellesmere summer, the sense that anything is possible for the explorer or photographer is enhanced by the unending daylight. The sun circles the horizon, illuminating one facet of a great rocky escarpment after another, until it comes around again to where the observer began to follow its journey. A photographer who is sufficiently patient is assured of front, side, and backlighting on almost any subject – a boulder, an iceberg, or an expanse of glacial scree. Hikers can begin a walk at midnight, at three o'clock in the morning, or at six in the evening without ever having to worry about returning to camp before dark. But the spring or autumn sun lingers just below the horizon, bathing the sky in hours of glowing sunset colours, a memorable experience for any intrepid hiker-photographer. At those times, icebergs, shimmering red in the polar sunrise or sunset, evoke a vision of Namib dunes that have been detached from the African desert and frozen for safe shipment to the other end of the world.

The ice fields and glaciers and summer's shallow ponds and bogs on the tundra make it difficult to think of Ellesmere as a desert – even though the Arctic air produces only a meagre amount of precipitation mainly due to the low level of evaporation in this cold region. Since only a shallow layer of surface soil thaws during the brief summer, and the subsoil is permanently frozen solid, the soil scarcely absorbs any water and plants cannot put down deep roots. A few trees exist, but their tiny contorted trunks look like roots lying on the ground – on the tundra it's not unusual to walk on top of a forest without realizing it. The boggy meadows support a complex of cotton grass, sedges, dwarf heaths, low herbs (such as the yellow Arctic poppy and creamy-white avens), mosses, and lichens with the vegetation often varying considerably between hummocks and depressions. However, like the Namib, much of Ellesmere is exposed rock desert buffeted by strong winds and abraded by blowing sand. Crustose and foliose lichens cover the rocks, except near the edges of retreating glaciers where the stone has only recently become exposed to light. However, in the shelter of cliffs, boulders, and depressions various heaths, louseworts, and clumps of rosey avens may form colourful gardens that follow the shapes of crevices, replicating the floral landscape of the ancient rocky hills of Namaqualand in southern Africa.

In both deserts, plants adapt their structures and habits to resist long periods of drought and to avoid being destroyed by wind. Many perennial plants in the Namib have tough, rubbery skins that prevent loss of moisture, and others, as on Ellesmere, keep their leaves and branches close to the ground, where the wind is less severe. Here snow and ice insulate roots and branches against the cold and provide a supply of meltwater during the short growing season; in the Namib, sand protects roots from searing heat. Many Namib plants also evade drought by existing as seeds during dry periods, but in the far north the climate does not guarantee a growing period long enough for most plants to set fruit, so Arctic plants propagate mainly by vegetative means.

Animals cope with the extremes of climate and weather on Ellesmere in many of the same ways that Namib animals do. Some birds and animals migrate; those that are incapable of long journeys may hibernate or simply reduce their level of activity. Many species burrow into snow or sand for protection. In both Ellesmere and the Namib, life cycles are speeded up. For example, many birds that nest on Ellesmere feed their young for several more hours a day than birds of the same species that nest farther south, so the fledglings are ready to leave their nests several days earlier than their southern cousins. In these and many other ways, desert animals adapt to their environments – and survive to propagate their species.

During a visit to Ellesmere at the height of summer, I made some close-up pictures of individual plants and animals, but concentrated mainly on the overall environment: a group of flowers blooming in the flowing meltwater at the tip of a glacier, or a small herd of musk-ox grubbing out an existence on the sparse vegetation of vast, rocky tablelands seemed to express more adequately the nature of life in these marginal habitats than did individual portraits. With a wide-angle lens, I could include in one image an Arctic poppy and the entire glacier behind it; with a medium-length telephoto lens, I was able to capture the musk-ox in the enormous grazing area they require; with a 28-85mm zoom lens, I could capture both the habitat and movement of Arctic terns that dive-bombed me when I intruded on the gravelly sweep of ocean beach where they were rearing their chicks. Sometimes I spent hours making compositions of colourful lichens that studded boulders and rock faces, now and then moving in close to feature a tiny section of lichen growth. One day at the start of a snowstorm, I moved from one spot of grass-and-heath-covered tundra to another, photographing the brown and green threads of grasses interwoven with white flakes and occasional swatches of reddish leaves and pinkish or creamy flowers. The tapestry kept changing, depending on the presence of sheltering rocks, the particular combinations of plants, and the accumulation of snow. The heavy overcast softened the tonal contrast between the snow, the darker earth and the plants, emphasizing the delicate beauty of the scene's overall fabric, while reminding me of the harsh conditions in which life is carried on. In all these situations, as in most others on Ellesmere, it was the environment, the grand overview, that gave meaning to a particular circumstance or event.

One day early in my trip I picked my way through a jumble of rocks up to a very high plateau that gave me a commanding view of the water and ice below, of the great escarpment wall that seemed to rise vertically from the distant shore, and of the ice field that lay on top like a vast down quilt. Far to the east, the mouth of the fiord opened into a larger and longer fiord, which in turn emptied into the ocean. I sat for a long time in the hard, crystal light, watching from the plateau like an eagle from a high perch, scrutinizing every shadow, studying every movement in the ice, cocking my head one way to catch the distant bellowing of a walrus, then another way to hear the siren whistle of air being sucked through a rock crevice, leaning forward to discern the strangely featureless, white sound of a glacial waterfall almost beyond hearing, and turning in fright when I mistook the tip of a cloud welling up from a deep valley for the white shape of a polar bear. It was as if I had to take Ellesmere into myself, to feel it, before I could begin to photograph. I had the time I needed, because today was also tomorrow, and the next day, and the day after that, and there would be no night or darkness to hurry me back to camp.

Eventually I felt sufficiently aware of the tones and hues, the textures, shapes, and perspectives of this great vista to begin photographing, so I elevated my tripod, attached my camera, and chose a lens. The sun had travelled nearly 45 degrees around its arc above the horizon since I had reached my vantage point, and now its rays were streaming across the far wall of the fiord, illuminating vertical ridges and casting dark parallel shadows between them. The basic pattern and its subtle variations formed the visual equivalent of a baroque concerto, demanding from me the most precise composition. I could feel, almost hear, the music, but to recreate its structure in the limited space of the viewfinder required patience and precise attention to detail. A black line placed too close to the left or right edge, or a highlight that distracted from the overall balance, had to be thoughtfully repositioned, along with the other elements in the picture space. Next, I focused on attenuated Z-patterns of open water zigzagging through the pack ice that glutted the nearer reaches of the fiord, taking care to vary my exposure as I zoomed closer to emphasize the darkish water, or as I zoomed back to include more of the ice. Finally, I aimed my lens at the mouth of the fiord where small icebergs and cakes of second-year ice sailed the ebony-blue sea in a midsummer regatta. Several dark brown, humpbacked islands, all of them rimmed with a continuous ruffle of shorefast ice, resembled gigantic puff pastries served up on a doily, or the drifting heads of Elizabethan noblemen. The effect was of a surrealistic theatre set. To augment that impression I added a polarizing filter to my lens, which delineated edges sharply and dramatically deepened the colour of the water that had been diluted by spectral light. However, the scattering of light rays by the atmosphere was less of a problem here than in the dunes, where the air picks up much more sand and dust when the wind blows with even moderate force and where, near the coast, the air may actually contain more water molecules than on Ellesmere. Previous experiences in the Arctic, the Namib, and other places where the air is often very clear and the light sharp, were sufficient warning to keep an ultraviolet or a warming filter on my lens at all times, except perhaps when a bluish cast was appropriate to the scene or my use of a particular kind of film provided the counterbalancing warmth.

After I had completed my photography of the earthscapes below me, I turned to the vast field of snow and ice covering the plateau behind me – and found myself squinting, even with dark glasses, at the reflected sunlight. At first, because the contrast between the ice of the glacier and dark promontories of rock was so extreme, so overwhelming, there seemed to be only two tones – white and black – and no hues whatever. However, as my eyes adjusted to the light I began to perceive some middle tones and gentle variations of blue and aquamarine. Then, gradually, I became aware of alternating grey, white, and sometimes pale blue lines, long parallel ridges and grooves that swept downwards from a light covering of fresh snow on the ice field to the very tip of the glacier. Created by flowing water, accumulated dust and dirt, pressures within the ice, the inexorable movement of the glacier, and other natural factors, these striations gave a roughness and distinct surface pattern to the glacier, a texture quite unlike the ripple patterns arranged by wind blowing on the Namib dunes (see the photograph on page 81). To produce on colour film the stunning whiteness apparent to the eye required approximately one and a half *f*/stops or shutter speeds of exposure beyond what a reflected-light meter indicated. Again the Namib dunes came to mind because, except for early and late in the day, they are lighter than middle tone and must also be overexposed (though not as much) if one wants to record accurately the contrasting tones and gently varying hues.

Moving along and down the glacier, I found myself walking beneath a second tongue of ice that had pushed out from the ice field through a high valley. On this ice face, which appeared almost as a vertical wall from my position, the ridges and grooves in the ice were enhanced by strong sidelighting. I stopped and set up my tripod, which I had been using as a kind of brake during my descent over the ice. Because the light was so intense, I was able to expose film at fast shutter speeds while using small lens openings for maximum depth of field, so the tripod was not necessary for sharp images. However, it was critical for good composition. The placement of the ice lines in the viewfinder was not the simple matter it seemed. Just as with the rock wall of the fiord I had photographed earlier, the position of an especially dark groove or an ice ridge brighter in tone than surrounding areas had a critical effect – good or bad – on the overall design.

Monochromatic, multitoned scenes are a common feature of the Arctic earthscape. Flying north from Resolute (Qausuittuq) across Cornwallis Island to Ellesmere, one sees Earth's brown, gravelly surface embroidered with black shadows and with lines and swirls of snow lingering on into the summer. The white ice mosaics of the ocean are often rimmed in sky blue or dotted with turquoise. And, it is only from close up that one perceives the multicoloured pointillism of the tundra created by the myriad hues of tiny flowers, mosses, lichens, and stones.

A single or generally pervasive hue combined with an extended range of tones often produces a starkly graphic simplicity that is both visually and emotionally appealing. The single colour establishes the mood while the tones play out the variations – reinforcing the theme and preventing monotony. Human injections of colour into the earthscape, such as a person wearing red or yellow clothing, become points of dominance that compete with the natural integration of the main hue and tones. Except for rare occasions when I have wanted to show the infrequent presence of humans on Ellesmere or to contrast the size of a person to that of a vast sweeping earthscape, I have avoided anything that interferes with the natural colour and tonal schemes that convey the sense of wilderness.

There is more to my restraint than a purely visual consideration. For me, it is a refreshing and humbling experience to be in places where our species does not dominate – where Earth's present condition and appearance is the natural expression of four and a half billion years of evolutionary development. Because there are so few places left where we have not obscured Earth's natural history by significantly altering its topography and plant and animal communities, those that remain are to be treasured. In environments like these, where the animals and plants take from the soil and water *only* what they need, and give back in equal measure, I feel more at home and in touch with myself than anywhere else on our planet. Thus, as a photographer in a wilderness area, I like to keep the human element in proper perspective, to show members of our own species as occasional migrants, and to emphasize the native species and topography.

When I visit a wilderness area I try not to let labelling or classifying the wildlife and habitat get in the way of pure observation, appreciation, and enjoyment. For instance, while I may spend some of my time identifying one plant as an "avens" and another as a "lousewort" and learning some of the ways they have adapted to the Ellesmere environment, I also take the time to enjoy the imperfections in the blossoms, the water drops on the leaves, and the hairs of a stem dramatically backlighted by the sun. Like making a friend, there is so much more to know than the person's name.

One morning as I walked along the shore of a fiord in a pale grey mist, huge cakes of "second-year" and "multiyear" ice stranded by a high tide attracted my attention. However, the labels seemed irrelevant to me. During the time I photographed the ice sculptures (see pages 76 and 77), my mind wandered freely over their forms and their reflections, observing, touching, listening, and feeling. By setting their names aside I had a more complete experience, not only of ice, but also of nature as a force that shapes and molds Earth's materials. I felt elated and liberated in the way one does when a great work of music engulfs and stirs the soul. Wilderness has this capacity for freeing us from our usual human concerns and worries that often inhibit positive emotional responses.

Even if we cannot visit wilderness, we can share in what it has to offer by learning about wild places and enjoying them through pictures, films, and books. Sometimes this is the best way to make a journey, because wilderness cannot survive the pressures created by large numbers of human visitors. At this time in Earth's history, we must regard a visit to Ellesmere, the Namib, or any other wilderness as a privilege, not as a right, and leave at home those attitudes and belongings that are harmful to its existence. In fact, there are times when we can express our caring for Earth more by the trips we do *not* make than by those we do.

In the forest on my land, there is a shaded glen where a periodic waterfall cascades over mossy rocks, providing moisture for an unusual variety of woodland flowers. This is an extremely fragile environment, except in winter when everything is frozen or protected by snow. During the other three seasons I permit myself only two visits, and once I have used up my visiting privileges for the year I look at photographs I have made there, and enjoy my memories.

The kinds of photographs we make in wilderness areas can often be recreated closer to home. For example, the photograph on page 175 was made on Ellesmere, but I have made others like it just outside my front door. In both places my attention was drawn to the melting edge of a crusty snowbank by the plinking and plonking of water drops, and both places offered me the opportunity to make an exciting visual journey, to explore the universe in a small space. For this kind of experience I use a macro lens or other close-up equipment for the closest focusing distance and minimum depth of field. By focusing on the edge of the melting snow, I am immediately transported to the realm of interstellar space. As I move the camera slightly forward, backward, or along the snowbank, stars and planets appear and disappear, whole galaxies rise above the horizon, followed by even more distant constellations. On other occasions, often without a camera, I explore prehistoric jungles in the mosses of my woods, hike across the South American pampas in the field behind my house, sail the Great Lakes on puddles in my driveway, explore vast deserts in a local gravel pit, and scale the Himalayas in a heap of rocks. From eye level I can fly over every conceivable sort of earthscape. I can see again the white-and-brown stretches of an Arctic island in a patch of snow-lined gravel, gaze down on desert dune patterns as I walk along a sand-rippled beach, observe the undulating sweeps of an ice field in any meadow after a snow-and-sleet storm, and witness the spring greening of the prairie on a bit of lawn. All these trips are free and require almost no advance planning.

One of the finest tributes we can pay to wilderness and its inhabitants is to bring them home with us, not as rock or plant souvenirs that deplete wilderness and gather dust on our shelves, but as living memories that we rekindle with imaginative seeing and photography wherever we live. In this way and in this spirit we can all visit Ellesmere or the Namib desert as often as we want.

The play of light and shadow dominates the Namib landscape, transforming natural forms and creating new, boldly graphic ones. Because the sun is always moving, the appearance of shapes, lines, and textures is forever changing. However, except at sunrise and sunset, the movement is slow enough to allow a photographer time to anticipate new configurations, and thus to choose camera positions and exposures carefully.

At sunrise, when red light and black shadows stream across the gigantic dunes of the Great Sand Sea, I often climb to a ridge of sand so I can see the sweeping shapes both above and below me. In this landscape, where all the familiar labels are missing and where there are few objects to provide scale, I often find it impossible to estimate distance. At sunset, when the black areas are red, and the red areas black, I still have no clues for establishing perspective.

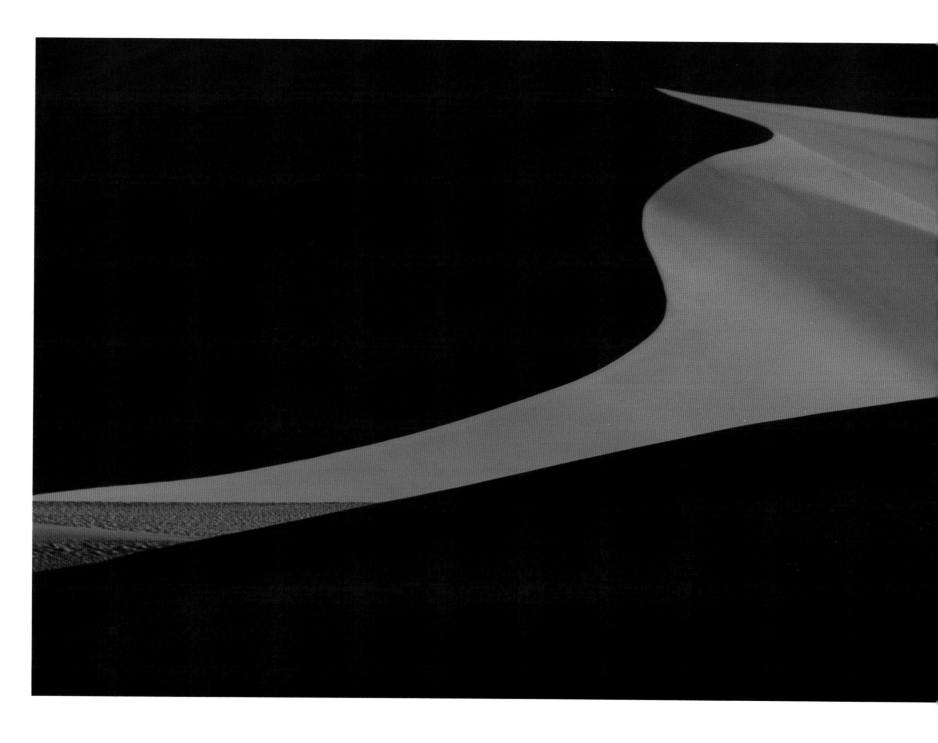

The dunes seem to glow most brilliantly when the sun rises or sets in a clear sky. Clouds or sand in the air scatter the light and soften the colours. So I study the horizon at first light or about an hour before sunset, and then head for a spot that seems likely to provide the best opportunities for photography – although one place is probably as good as another, each simply offering different opportunities. From the spot where I made this picture, the graphic designs changed so constantly that I took a day's pictures in one hour.

There are trees in the Great Sand Sea, magnificent trees. They grow in river beds, where their roots reach deep into the sand for residual moisture, because there may be no water in the river for years on end. These trees provide food for birds and insects, as well as nesting sites. The insects, in turn, become food for birds, lizards, and other insects. Large mammals, such as gemsbok and humans, use the trees for shade, and smaller mammals may make their homes among the roots.

After making the photograph on the facing page, I walked just a little past the tree to make this picture of a dune rising behind it. Away from the river bed, there often seems to be no evidence of plant life, yet when I climbed the dune I found seeds and old dry plants that had been scattered across its surface by the wind. When the wind blows from the west it carries seeds a long distance inland; then an occasional easterly gale carries sand in the opposite direction, burying the seeds, where they wait for sufficient moisture to germinate.

OVERLEAF Backlighting changes the colour of these dunes from brick red to almost white.

One afternoon I made a difficult climb through soft sand up a steep dune. From my vantage point on the ridge, I could see that, as the sun neared the horizon, it would probably throw most of the scene below me into shadow, while casting a golden band of light across the more distant sand. So I set up my tripod and camera, made trial compositions, then waited for the most graphically expressive moment.

In the afternoon light the ridge of the dune and its shadow seem to point down at the tree that the sand will soon engulf. These dunes advance slowly, but relentlessly; they have been increasing in size and moving across the desert for two to three million years. The fate of this tree is inevitable.

One afternoon, with friends, I entered a very strange world. Hundreds of years ago, advancing dunes dammed a river that flowed periodically to the sea. The trees that lived in the river bed probably survived for a long time, then gradually they too were submerged by the sand, which dried and baked them. Eventually, the winds uncovered the dead forest. However, many more trees may still remain buried in the surrounding dunes.

Another day I returned alone to this eerie place. For awhile, I photographed from the vantage point of a dune. Then, I walked down to the trees and the wide mosaic of white mud tiles. I sat under the skeleton of a tree and listened to the pure silence of this haunting wilderness, a most uncommon experience in a world of cars, airplanes, radios, and television.

One day when I was hiking alone, I found myself in unusual surroundings. Although it was late morning, the sand glowed a vibrant pinkish hue that varied through a range of lights and darks normally seen only at sunrise and sunset. The edges of some sand forms were precisely defined, while others were not. The side of a dune was beginning to collapse. A lone gemsbok had walked through without stopping. Something seemed strangely wrong and uncanny. Only by making photographs was I able to quell my rising sense of fear.

The skeletal remains of camel thorn trees are found in many parts of the Great Sand Sea. Here the ripples of sand seem to take on the form of their roots. I used a wide-angle lens pointed down at the ripples to show the effect. Perhaps one day the wind will blow the sand away, revealing the underlying mud tiles, or, perhaps the people who drive their cars from the nearby road onto the sand will scar the face of the land forever.

After eleven years of drought, sufficient rain fell in the highlands to the east of the Great Sand Sea to send water coursing through this river bed. The water collected in a lake behind a dam formed by the dunes. Six months earlier I had hiked across the dry mud bottom of this lake, never expecting to photograph one day the dunes reflected in water.

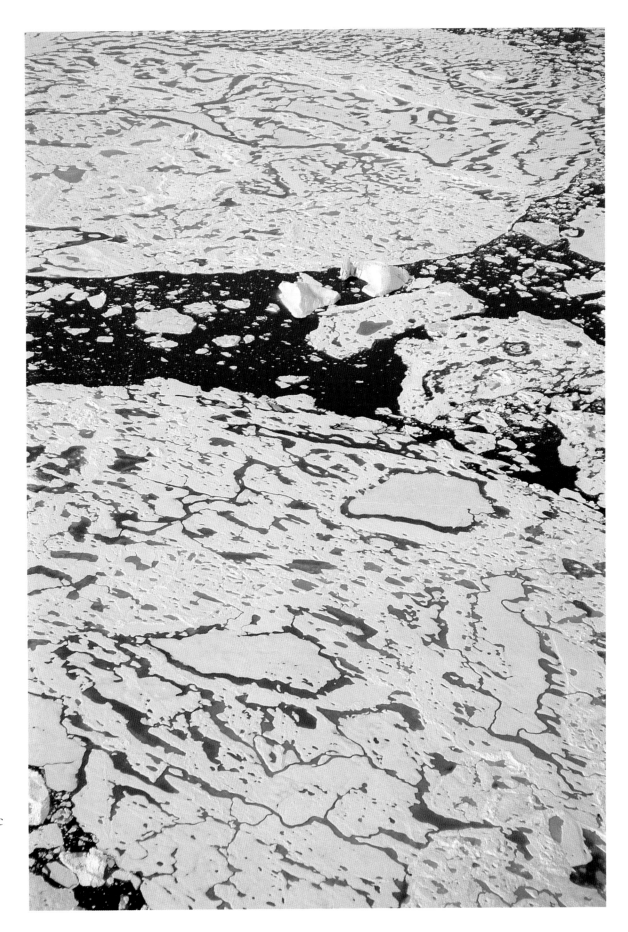

Viewed from the air, ice floes in the Arctic Ocean often remind me of mud-tile patterns in the Namib Desert seen from my normal eye level. I am constantly struck by the visual similarities of these extremely different environments.

Much of the Arctic's water supply exists as ice and snow in forms that are often quite remarkable in both design and colour. Ice that is more than a year old takes on a luminescent blue cast that is especially beautiful in the soft light of an overcast day.

Summer heat and wave action shape the huge ice cakes that sail the Arctic seas. Stranded by a high tide, this gigantic ice sculpture sat for days on the shore of an Ellesmere fiord. Hopping from sand bar to sand bar at low tide, I managed to find a position where I could show its striking design.

The reflections of ice sculptures are as intriguing as the sculptures themselves, especially when they are viewed separately or form a backdrop for other ice cakes. I tried to see the various ice forms as things-in-themselves, new realities for the eye and the mind to explore.

For most people, seeing without labelling is easiest with new objects and new designs. We have fewer visual preconceptions then, and are more alert to shapes, lines, textures, and hues. This heightened perception helps us to respond more freely to the subject matter. However, with a little practice, we can learn to see familiar objects and designs in the same way.

Perhaps the most striking visual difference between the Arctic glaciers and the Namib dunes is the range of tones. There are deep blacks and middle tones in the Great Sand Sea, but few highlights. The ice fields of Ellesmere Island, on the other hand, seem to offer every tone from brilliant white to black. In fact, colour often plays a secondary role here except at those times of year when the sun is very near the horizon and adds colour to the earthscapes.

In summer the glaciers provide meltwater, which collects on the surface because the ground is permanently frozen not far below. The Arctic plants depend on this surface water for growth and spread their roots in the shallow layer of soil above the permafrost.

This picture was taken from the top of an escarpment a great distance from the scene below. In the Arctic, as in the Namib, distances are vast with little to provide scale. Here, one can see the characteristic features of Ellesmere Island – the ocean packed with ice, the brown, treeless tundra, and the exposed rock scree at the base of a glacier.

From another escarpment, I gazed down on ice breaking up in a fiord and being blown out to sea. Because the wind was strong, the ice detached itself from the land and moved away rapidly, forcing me to alter my composition frequently. Even though I used a long telephoto lens to isolate the zebra-like ice pattern, the photograph covers a broad expanse of water.

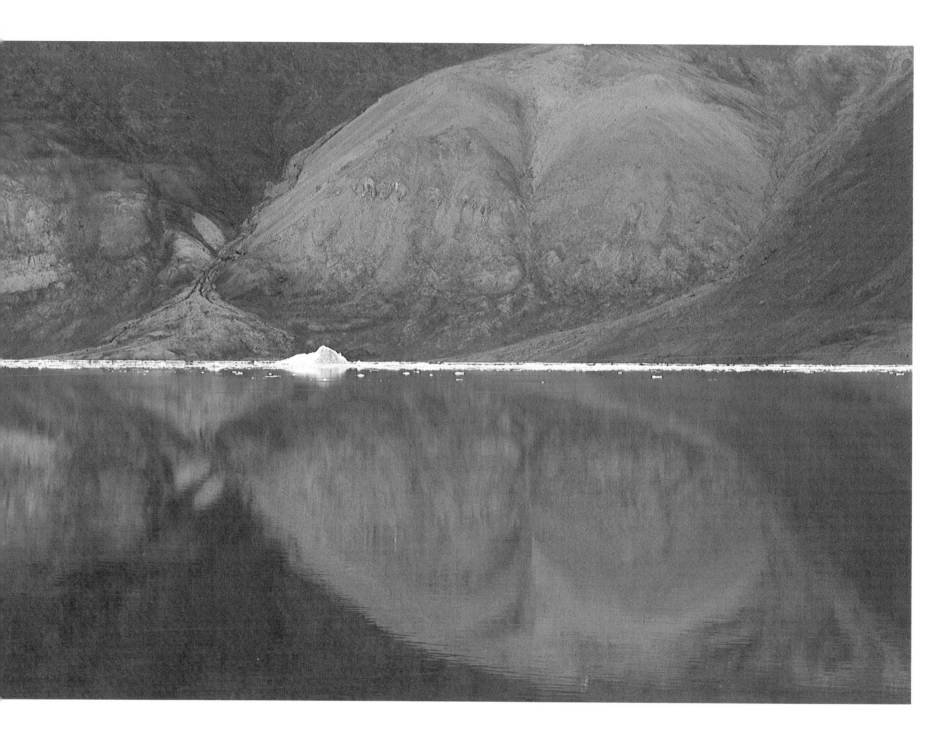

Despite the harsh climate, summer on Ellesmere brings moments of serenity. Perhaps one treasures them more because of their infrequency and brevity. An hour after I made this photograph, the sky was overcast and the wind was whipping snow across the water. The next morning, huge cakes of old ice clogged the fiord and new ice covered the water between them.

PORTRAITS OF HOME

Take care of the land, and it will take care of you.
Take what you need from the land, but need what you take.

ABORIGINAL LAW

The Forest

I live on the edge of a forest; the trees begin moments from my door. I know the nearer parts so well that, when I am away from home, I can wander through the forest in my mind, stopping here to examine a delicate starflower in the spring, pausing there to watch winter's first snowflakes gather on the thin ice of a leaf-choked puddle. I go from place to place and season to season – listening, looking, smelling – until the forest in my imagination seems as full of birds and ferns and as fragrant with cedars as the real forest I left behind.

When I return home, the first walk my dog and I make together is into the woods. This is as special for Tosca as it is for me, because she does not venture far into the forest by herself. It is not her territory. We enter the forest by an old logging trail that begins just behind my woodpile – Tosca out in front, ears and tail alert. She will discover far more on the walk than I.

The trail is straight at first and level, except for some deep ruts gouged by the wheels of my tractor one warm day in late winter when the frost in the soil melted before I had hauled the last load of firewood out to the meadow. I try to avoid making such scars, by using the tractor only when the woods road is frozen, and I'm happy to see the ruts filling up with maple leaves and other organic debris. On either side of the trail, among the young birches, maples, red osiers, and aspens, hay-scented ferns form a lush, deep carpet – green in summer, brown in autumn – that obscures the soil. Sometimes I find ghostly Indian pipes and poisonous Amanita mushrooms beneath the fronds that lean out over the trail, and I photograph them by reflecting light into their dark spaces.

Farther along, Tosca sniffs at a favourite old stump. Her nose tells her that racoons live here, but I can only tell by the droppings. It is a good stump for small animals, partly uprooted and easy to hide under, and neither of us has ever seen the racoons. Then the path becomes wet with the spill-over from a bog. In winter when the bog is frozen, huge hummocks of cinnamon fern roots project through the ice like mountain peaks through a layer of clouds. In summer, the tall and stately fronds wave slowly in the breeze; there is never much wind here, because of the surrounding trees. In the centre of the bog stands a magnificent maple, the parent of ten thousand young maples in the woods around, and a favourite hunting perch for a pair of barred owls. Once, when I aimed a telephoto lens at them, they posed for me as if they had been expecting a sitting. Also, I have found moose tracks here on numerous occasions, and heard the great beast crashing through the alders on the far side. It visits only periodically, because a moose needs many bogs for browsing; and despite its massive size and huge appetite, it leaves the place pretty much as it found it, with only some willow branches missing and some grasses nibbled back below the waterline.

Beyond the bog the trail inclines upward to drier ground. I can see far into the woods on either side, except for mornings when mist drifts through the trees. To the left are cedars, spruces, and birches, with an occasional bilberry, beech, or aspen and a few thick clumps of young firs. To the right the land rises sharply to a high rocky ridge and the hillside is an open forest of tall, slim, white birch trees lightly shading a thick ground-cover of mosses, gold threads, wild lilies of the valley, and bunchberry blossoms, punctuated here and there by pink lady's-slippers, creamy pyrolas, and just possibly by striped and spotted coral root. Sometimes I lie on the moss, gazing up at the weave of leaves and branches, wondering what sort of lighting will enable me to show this natural design most effectively against the background of sky. If I remain there for a long time, Tosca will return from her explorations and lie down quietly nearby – whether out of affection for me or a sense of duty, I do not know.

It is easy to walk through this part of the forest, because a neighbour and I have done extensive selective cutting. Years ago, a previous owner clearcut the area for pasture, and it grew back into a forest that bore little resemblance to the original mix of deciduous trees and conifers. The few living trees were all of the same age and the spaces between them were clogged with an abnormally large number of bushes and smaller trees that had failed to survive the merciless competition for light and space. An unnatural mess! So, we left the living and took out the dead, and here and there planted young trees of several native species. In time, it will be a proper natural community again – a diverse habitat for animals and plants of all ages. The abundance of low plants thriving on the forest floor indicates a good beginning, and the ruffed grouse have already moved back. In winter they nibble on the buds of the younger birches, and some have nested in the ferns that grow around piles of dead branches that I left behind as protective cover for birds and small animals. However, deer still avoid the place and hare, which are abundant everywhere else, seldom visit, because there is as yet an insufficient growth of young evergreens to provide protection from storms and predators. I am photographing this section of the forest every year in order to record its progress back to full health. In my overall, medium-distance, and close-up images I record the annual changes in a few specific spots, but for the project to have real documentary value, I must continue it for many years.

The trail curves around the ridge and angles steeply down a long and sometimes precipitous slope to water far below. The vegetation changes here, because the ground is loose and rocky and the hillside faces southeast. A few white pines, the largest trees in the forest, tower over clusters of Christmas ferns and groves of young beeches. A dead giant, precariously near to toppling over, provides a home for a host of insect species and endless meals for flickers and downy woodpeckers. Pine seedlings dot the forest floor because the old mammoth has lost its needles, allowing sunlight to flood into the lowest stratum of the woods. Many of them will be crushed when the big tree comes crashing down, leaving more light and space for the survivors; and the huge log will lie there for years to come, serving the wild community in more ways than I can imagine. Perhaps I should start now, before it falls, to record the story of its life after death.

The trail grows steeper. Tosca and I descend through scatterings of tiny, fragrant white violets in the spring (see page 106), or through patches of blue columbine when summer has arrived. A yellow mullein, taller than I am, grows in the middle of the path – another reason for using my tractor on this trail only in the winter. Here the forest is dense, the thick stands of evergreens providing excellent shelter for animals and birds. If I lift a low-hanging spruce bough, and peer into this dimly-lighted realm, I can see ten thousand footprints in the snow. It is a busy world in there, as winter's tracks remind me.

The logging trail splits abruptly at the edge of the creek, which is actually a long, broad river inlet. If I turn to the left, I enter one of my favourite places on Earth. The soil is carpeted with twigs that have fallen from immense cedars that line both sides of the trail. It is like walking up the aisle of a cathedral. My father once suggested I should harvest the wood before the trees became rotten in their cores, but I decided that most of the trees would outlive me, and if I were to cut them I would not see their like again. So they stand there in the hush of early winter, snow hanging lightly on their branches; and they are there in the spring when the creek rises, flooding the trail and making it possible for me to paddle my canoe down an avenue of water. In summer the trail beneath the cedars sometimes has lines of tiny, white, capless mushrooms that grow nowhere else that I know of. They seem to mark the farthest advance of the spring flood waters, and I wonder about the conditions that create such a narrow, specialized habitat. In autumn bare spots that have appeared during the year in the carpet of twigs are rewoven, either by squirrels who nip off twigs from high branches and forget to collect those that have fallen to the ground, or by the trees themselves as they discard old vegetative material.

If I walk to the right along the shore, the trail soon brings me to a stream that tumbles through an open beech-and-aspen wood down the rocky face of the escarpment. At its base, in the thick mud near the creek's edge, grow ostrich ferns, masses of them. In early May some neighbours come to pick the emerging croziers, called fiddleheads, which are delicious when lightly steamed and served with butter. Nearby the large, heavily-veined leaves of false hellebore are also unfolding. It is easy to while away a spring morning here, especially a cloudy one, when the new greens seem to glow in the soft light. At this time of year, I usually bring along my tripod, camera, and macro lens to photograph the linear designs of the false hellebore leaves and the unfolding curls of the fern fronds. Sometimes I concentrate on the meaning of the natural designs – the reasons for the long, nearly parallel lines in the false hellebore leaves, for example. Other times I abandon myself to experiments in abstract visual design, perhaps varying both camera position and depth of field from one composition to the next. For both kinds of photographs the physical appearance of the leaves serve as a starting point. Tosca is extremely patient with me in all of this, but sometimes she falls asleep (both ears up) on a mound of old grass. Once, in a careless moment, she flopped down on a grassy anthill, which was drier than the surrounding soil. But she didn't remain there long!

I especially like to climb the stream bed after a cold snap in late autumn or early winter, before any snow has fallen. Then the waterfalls are shimmering stalactites, the branches of fallen logs are sheathed in ice, and the blackish-green mosses enshrouded in rime. In the afternoon, when the hillside is in shadow, the contrast between the highlights in the ice and the blacks of the mosses is sufficiently reduced to allow me to capture detail in both areas.

At the top of the escarpment, where the water flows out of an extensive bog and first cascades over the rocks and rotting logs, I always pause to catch my breath. In early winter it is a hard and slippery climb to the top, even when I am in the best of physical condition, but there is another reason to rest: the grey ice surface of the bog swirls with patterns of silver-white lines. These intriguing manifestations of the relationship between flowing water and freezing temperatures are simply too good to pass by without a long and appreciative look, and sometimes, pictures.

Tosca senses that we are now on the return portion of our journey, but waits for a signal from me before bounding off through the bushes above the bog. She knows exactly where to go, but so do I. Her nose smells one set of clues; my eyes read another. So we set off for home, two creatures who share the same environment, yet live to a large extent in different worlds. I hear her barking far ahead of me, and wonder what danger she has encountered. When I catch up with her, I discover it is only a grey squirrel – but it might have been the moose, or a bear.

After Tosca has chased away the squirrel, we enter the final stretch of woods. While it is quite still here on the forest floor, the lowest level of a forest being the most protected from the effects of weather, the highest branches of the hemlocks and spruces are bending and tossing. The treetops tell me that the wind is blowing from the southeast, threatening rain or maybe snow. Perhaps the next storm will finally topple the old pine tree, but perhaps not. I have seen trees withstand a hurricane, only to collapse a few weeks later in a moderate breeze.

I come at last to the mossy rails of a cedar fence that was erected a century ago. It marks the boundary of the old pasture, and I know I am nearly home. Tosca is not here; she will be waiting at the edge of the woods. I jump the low rails back into the section of forest that has required so much selective cutting and replanting, and notice in the gathering darkness that it is becoming wild again. One day all the evidence of my labours will disappear, perhaps not in my lifetime, but soon. And with that thought, which has secretly pleased me countless times, I join Tosca for the walk across the meadow to the house.

Waterscapes

Water is a dynamic and expressive medium, constantly changing its image and its voice. Think of the roar of the surf, the smashing of waves, the thunder of a cataract, the raging of a flood, the ominous cracking of ice in a tidal river, or the quieter lapping of ripples, the gurgling of a small brook, the tinkling and plinking of falling drops and, if we listen very, very carefully, the barely audible chorus of "pings" and "pops" emanating from a melting snowbank.

Water is a more constantly active and noisy medium than soil, which occasionally erupts in a volcanic fury or shifts periodically along fault lines, but usually remains relatively silent, keeping its deeper secrets to itself. As a liquid, water moves easily, or flows, adapting itself readily to the forms of the land. As an invisible vapour it rises, then condenses into clouds, mists, and fog, which appear to be always moving and changing. Even as ice, frost, and snow, water displays an impermanency, a capacity for change that is rapid compared with the generally imperceptible evolution of land or soil.

This *natural* dynamic of water has a powerful impact on the human psyche. We can relate to water's moods. In fact, we perceive important similarities between the behaviour of water and the behaviour of people, and often describe ourselves with words we use for water – "he babbled on interminably," "his rippled brow revealed a sea of cares," "she spilled out her troubles," "still waters run deep," and "a flood of emotions." We speak of a person with an open and active mind as having "a fluid imagination," probably connoting the dynamic of water. On the other hand, we often regard land or soil (especially rocks) as a symbol of permanence, and we speak about "landforms" (implying fixed designs) rather than of the dynamic of the land.

Water has a *graphic* dynamic as well. Its capacity for reflecting and blocking light is such an important characteristic that we can think of water surfaces as being constantly painted and repainted by light. The appearance of water is influenced more readily, more rapidly, and to a far greater degree by changes in the colour, intensity, quality, and direction of light than is the appearance of land or soil. However, in this respect too, we humans are like water – our moods and activities constantly being altered by changes in light.

Thus, we photograph water not only because of a purely scientific or aesthetic interest in its physical properties and characteristics, but also because we are interested in photographing things that remind us, somehow, of ourselves. I have long observed that, in an earthscape containing a body of water, photographers will make more pictures of the water than of the land, and they will make the water images first. This happens so consistently that I have come to regard the perceptible dynamic of water as being a compelling psychological force in human vision. It is more profound than a purely physical excitement of the eye, which by itself would be insufficient cause for regarding water as a metaphor for human activity and behaviour.

When we make photographs, we control the visual effect of water's dynamic in two basic ways – through visual design and through camera technique. Let's imagine a typical mountain scene – a range of snow-covered peaks from which a creek of meltwater rushes into a valley below. The natural dynamic of the water is restrained by the edges of the creek – that is, the creek's shape is defined by the land on either side. By moving camera positions and changing lenses we can alter this natural shape endlessly – perhaps making it seem longer and more attenuated, or shorter and wider. By the same methods we can change the size of the creek making it tiny in relation to the mountain peaks, or seem to overwhelm them as the water spreads across the entire foreground. If we take a position that shows a tree intersecting the creek, we create two shapes; and again we can vary their relative size and importance. Or we may cut the shape of the creek into several shapes. Depending on our mobility and our range of lenses, anything is possible. In addition, we can vary shutter speeds to render the water sharp, textured and full of detail, or blurred. By making these and other choices that alter the appearance of the creek, we control the psychological impact of the water – its effect on viewers' thoughts and emotions. The basic question, of course, is *why* we make the choices that show the water in one way and not in another, and to that there are as many answers as there are photographers, and their moods and ideas. In the final analysis each photograph is not of water in a mountain scene, but of each photographer's response to this subject matter.

My most familiar waterscape is my bathtub. When it is filled to the usual level, my knees protrude through the water surface like two large triangular rocks, and an old brown bathsponge drifts like a mass of floating seaweed, while I observe everything from a position halfway up the island of my head. From here I gaze across a seemingly enormous expanse of water, watching as the visible dynamic of its surface changes with my activity, from placid to gently rippled, rough and textured, or turbulent. Waves move in whatever direction I choose, whipping up foam that swirls around the islands to faraway coasts. Reflections vary with the calmness of the surface, the amount of light streaming through the window, and the nearby arrangement of coloured bath towels, which occasionally appear as the warm rays of a sunrise or sunset on a glassy sea.

While I have made photographs in my tub on a few occasions, more often I regard my bath as an opportunity to observe the ever-changing appearance of water and to imagine the pictures I might make of an ocean, river, or lake. Sometimes when I rise before dawn, I lie in my bathtub until first light noting the ways the water surface resembles the kaleidoscopic appearance of channels, bays, and inlets of the great tidal river, the Saint John, that flows below my house. Then I head down through the fields to my canoe, hidden among the trees at the river's edge. Here is the waterscape I know second best. Wind seldom disturbs the water surface in early morning, and often I create the first undulations when I poke the nose of my canoe into the water. With a few easy strokes of the paddle, I glide across the mirror calm of the narrow inlet in advance of the ever-broadening wedge of ripples that follows the canoe. In spring, when the swollen river inundates sandy shores and marshes and partly submerges maples, aspens, and ash trees that grow along its banks, I stay close to the land, often heading through openings between branches into temporary coves and bays where I can observe a familiar world from an unfamiliar point of view. Beavers and muskrats occasionally swim by on the night's final errands, and small flocks of migrating Canada geese move from the water into fields that last year were planted in barley and oats. As the day brightens, the water mirrors broad bands of new green leaves emerging in profusion on the aspens (see page 105), and when I pause to lift my camera to my eye, the ripples surge past the canoe, then weaken to a vague disturbance as I drift into motionlessness. Less than an hour earlier, in my tub, I watched ripples dissipating to an ''almost calm'' in the same way as on the river surface. With the gentlest of movements, I begin to photograph.

First, I concentrate on the overall scene with the reflection of the budding aspens occupying the lower half of the picture space. Then, I focus on the reflections alone, while holding my breath to still the canoe and the camera in an effort to capture the mirror-like image. Finally, I begin to rock the canoe slightly, so the gentle greens and greys of the reflected trees become dancing lines of colour and tone. The rapidly changing surface makes every image significantly different from all the others, and each contains an element of chance, because I cannot press the shutter release quickly enough to capture exactly what I've seen before it changes. Just as I am about to paddle on, a pair of ducks splash down into the water and head immediately toward a small island of willow bushes. Their wake catches the first gentle sparkles of the sun, which is barely visible through the overcast sky. Switching to a telephoto zoom lens, I place their tiny forms in the upper left corner of the picture area, with their wake streaming out horizontally behind them and out-of-focus reflections forming nebulous colour patterns in the lower part of the space. When I press the shutter release, they are startled by the sound and begin to swim furiously for the safety of the willow bushes. Wondering if they will nest there, I make a mental note to paddle by when the ducklings have hatched, and perhaps to photograph the whole family out for a swim – if I'm lucky.

On windy summer evenings I battle waves that threaten to toss the canoe and me against the cliffs of a rocky bluff before I round the headland into the calmer waters of another inlet. Here, protected from the wind, I skim along in the shadows as the late sun warms the rich greens of the forest on the far shore. Three or four times every summer an unexpected thunderstorm suddenly transforms the sky, land, and water, and I head for a tiny alcove of sand protected by overhanging rocks and trees, where I can watch and photograph more easily and safely. Before the clouds obliterate the sun, there are dramatic moments when the sky and the water become dark and ominous, while the land remains brightly illuminated, glowing with an intensity of colour (see page 107). From my sheltered location, I point my lens at the clouds to obtain a meter reading, underexpose one or two f/stops to retain their dark tones, and then tilt the camera down sufficiently to include the band of forested hills and a strip of water.

Sometimes concentric ripple patterns created by raindrops dot the water surface before the wind strikes. Unless the last vestiges of sunlight and clear sky highlight the patterns, the tiny rings-within-rings are barely perceptible variations of grey, beautiful to observe but difficult to photograph close up in low light. (However, by hanging soaking-wet towels on the rack above my tub to drip, I discovered that I could make photographs of somewhat larger areas of raindrop patterns at a shutter speed of 1/30 second, or even slower speeds if I enlarged the area more by either backing off a little or switching to a lens of wider angle.) Other times the storm arrives as a wall of violently churning wind and water. One moment everything is calm and hushed; the next, leaves are being ripped from branches, waves are battering the beach, and every exposed thing is being drenched. The noise is always a surprise – tons of water being dumped from a great height, a veritable Niagara. If a burst of sunlight backlights the falling rain, I use 1/30 to 1/8 second shutter speed to produce long, slanting streaks, always including large areas of dark tone as background and underexposing one or two f/stops to make the streaks stand out clearly. At this time of day from my vantage point, frontlighting is more likely, and if a rainbow appears I will also underexpose one f/stop to keep the tones of the sky dark and the rainbow colours saturated. When the storm passes, I may paddle back around the headland and home, but often I leave my canoe at the end of a forest trail that leads up the back way to my house at the top of the bluff. Then I can return easily to paddle the same waters the following morning.

After a cold front has moved by, the sun often rises through pink or golden mist – a time of delicate mysteries. Through the early morning fog I hear the cries of unseen loons on the water and the songs of warblers and sparrows in the forest. Setting my camera for about one f/stop of overexposure to preserve the brightness of the mist, I may photograph an island of pines that suddenly appears high above me, then disappears almost as quickly, or I may use my telephoto zoom lens to move in on ducks feeding among grasses near a hazy shoreline. In shallow water at the end of the inlet behind my house, large patches of white water lilies alternate with stands of purple-flowered pickerel weeds. Winding channels between the patches form the paths of this huge aquatic garden. When a breeze begins to blow, the floating flower beds interrupt developing waves and the water of the garden remains placid much longer than it does in open stretches. As the sun shines more brightly and the mist gradually dissipates, I leave the edges of the garden where I have been photographing the leaves of arrowhead lilies silhouetted against the pale gold water, and navigate one twisting channel after another. Because I am so close to the water surface, my 20mm or 35mm wide-angle lens enables me to move in closely on a clump of water lilies or an individual pickerel weed blossom while showing an expanse of the habitat at the same time. It may be noon before I paddle out again and point the canoe toward the forest trail that I left hours earlier.

In the spangled autumn, brooks deposit their collection of fallen leaves into the dark waters of river bays. This is a time to walk the shores, to observe the leaves tracing out the movement of currents and eddies, and perhaps to photograph them (as I did, using a time exposure of four seconds for the picture on page 122). The water seems darkest at this time of year, as if it were closing its curtains against the coming winter. I peer into the blackness and see nothing – a night without stars. Even my reflection is sometimes strangely absent. Arching brown grasses disturb the surface with their dangling tips – in and out like a swimmer's toe testing the water. Sunlight skips and bounces on the surface with the alacrity of a hurled pebble. Nothing seems to be absorbed, except perhaps an old log that started to sink in the spring and spent the summer going under. On cold mornings the water's edge is fringed with translucent white ice, a thin and glassy version of the thick shorefast ice that clasps Arctic coastlines even during the height of summer. The wind and waves toss it onto the beach for easy melting by the sun, and soon this visual harbinger of winter is no more than a memory that one can put aside for a time. Then, one morning, the shore ice does not melt.

In the winter my morning bath provides me with few visual equivalents of the frozen river. Yet, during this season the natural and graphic designs of the river surface vary more dramatically than at any other time of year. However, since they change more slowly, they can be held by the eye, studied, and remembered in a way that the designs of water cannot. Early in the winter, before the snow falls, the ice appears grey from a distance, and slowly deepens to black as one cold night follows on the heels of another. But, close up, the black surface is like the Milky Way. First I explore on foot, making pictures of abstract patterns in the ice, always paying careful attention to exposure, which I must vary according to the number of white bubbles and cracks that dot and crisscross the dark surface (see page 132). Then, donning my speed skates and keeping my head down to avoid seeing the land, I glide away through interstellar space. Moons rise and set, stars appear and disappear, entire galaxies flash by. Sometimes I wear a portable cassette player and earphones, and listen to electronic music written for space odysseys as I sail on and out through the constellations; but at other times, when the tide is rising under the ice, the snaps, cracks, and rolling thunder of the ice are more than sufficient to engage my imagination.

Everything changes when the snow begins to fall. At first the wind chases the gathering flakes across the ice until they swirl into the lee of bumps and ridges, where they settle to form ever-expanding white patches and stripes. For a while, viewed from a hill or an airplane, the river surface resembles a zebra's coat. As the storm continues, the alternating patterns of light and dark disappear in a white mass, and soon trees on the shore become the clearest indicator of where the land and river meet. However, they are not the only clue. In tidal rivers, the ice is lifted by the water every twelve hours and then dropped. As a result, long, often continuous cracks appear close to the shore and water gushes up through them, flooding the surrounding ice and seeping through the snow. These elongated, yellowish stains persist through the coldest weather, their colour refreshed by every high tide, often providing the only hue in a world of white.

During a warm spell the broadest and deepest part of the river, where the current flows most strongly, opens up. Now great shafts of dark blue water slice through the snow and ice, softening in later afternoon to pale greenish-gold and eventually to gentle orange or pink. If a winter gale blows for even a few hours, more of the ice breaks up, and a lake, as it were, appears in the broadest and deepest part of the river. Sometimes towards evening I set up my tripod, camera, and longest lens on my front deck and make pictures of the bands of colour (which vary in both saturation and tone) as they lie one above the other in an illusion of two-dimensional space. Other times I watch from a window as evening approaches, and observe the water changing from pink to mauve as the surrounding snow and ice deepen to a rich, glowing blue. If a light is turned on in a farmhouse across the river, the spot of warmth becomes an accent in the twilight. As spring approaches, other lakes begin to appear in the river; and then, perhaps one windy night, the real break-up occurs. By morning, islands of ice are sailing away to the ocean, and I am heading down through the fields again to put my canoe back into the water.

One day I will paddle the Saint John River as far as the ocean, which is less than an hour's drive away. There, in the Bay of Fundy, the long arm of the Atlantic that separates New Brunswick from Nova Scotia, the tides rise higher than anywhere else on our planet. Twice every day the gravitational pull of the moon and sun causes a narrow swelling of water to sweep around the earth like the hand of a clock. Due to land configurations along the coast and below the water, this swelling tide is amplified as it enters the Bay of Fundy and becomes progressively higher as the bay narrows, cresting at approximately fifteen metres (fifty feet) above low tide. But only six hours and thirteen minutes later, water that lapped against high cliff faces and inundated salt marshes has receded and now can be seen only dimly across enormous expanses of sand and mud. It is a tidal spectacle unrivalled on Earth.

For me low tide is the most interesting time, because then I can hike on the bottom of the sea. Walking barefoot or in rubber boots as the tide recedes, tripod and camera tucked under one arm, I may head first to a sandstone bank to make abstract studies of the sculpted red forms, accented below the tide line by green and brown seaweeds. Or, my walk may be interrupted by patterns of sand ripples alternating with lines of tiny, storm-tossed clam shells. Or, I may follow a tidal drainage stream that curves repeatedly across the mud flats (see page 35) and, finding myself far from my original destination, begin to wonder when the returning tide will catch up with me.

As I slog through the mud, I am aware that here is one of the richest habitats on

Earth. Microscopic single-celled plants cover the mud like a living skin, producing their food at low tide when they are exposed to sunlight, and then becoming food for a host of higher creatures in the life chain when the tide comes back in. Snails and shrimp are abundant here, as are tube worms, clams, and many other forms of marine life. As the tide creeps in again, shad, flounder, and cod come to dine, swimming about where I was walking only a short time ago. During migration periods, shore birds that have scattered across the mud at low tide now gather in huge flocks on the narrow beach – sanderlings, plovers, and sandpipers, often by the millions. They soar and whirl in unison at the slightest disturbance – an unforgettable experience at sunset when the warm evening light flashes off a hundred thousand wings. A high-speed film is necessary to arrest the birds in flight, but a slow-speed film enables me to pan my camera or to blur the birds with a slow shutter speed, and thus to convey the sweeping movements of the flocks.

In other places along the Bay of Fundy, huge basalt columns line the shore, remnants of ancient lava flows. If the tide is low when these cliffs are in shadow, I may scramble along the rocky beach to capture the delicate hues reflected onto them from the water (see page 161), or to photograph hanging gardens of wildflowers that have established themselves in crevices. In the intertidal zone at the base of the columns, white barnacles stud the eroding rock. Like so many creatures that live in this turbulent environment, they have evolved mechanisms or structures for holding fast against the pounding waves. Here, as everywhere else along the Fundy coast, water is the dominant force, a megapower that nobody argues with.

People often find their first visit to an ocean an awesome experience. The sheer magnitude of the energy is beyond what we imagine. Long lines of surf race relentlessly to shore, fragmenting against rocks in explosions of white mist that boom like thunder through the continually pervading roar. Between explosions the water pulls back, sucking and dragging with it everything that is not fastened down, only to smash them a moment later on the rocks or to swallow them forever. Neither words nor photographs can ever adequately convey this experience to somebody who has never been there; there is no verbal or visual equivalent of an ocean's power. Yet, only a short distance out from shore, the effect is muted by the absence of land, and the ocean seems more a mysterious, brooding force.

Sometimes when we are on or beside the ocean, a dramatic burst of sunlight will illuminate part of the dark and turbulent water surface, evoking a sense of "being present at the Creation." We experience or feel the creative event because of the expressive contrast of tones – darkness suggesting non-existence and light symbolizing life. Perhaps this symbolism has something to do with our fascination for the sea. The brooding ocean represents the potential dynamic of the universe, and the light shining on it the actual dynamic of creation and the evolutionary process that continues to shape Earth and mold our lives.

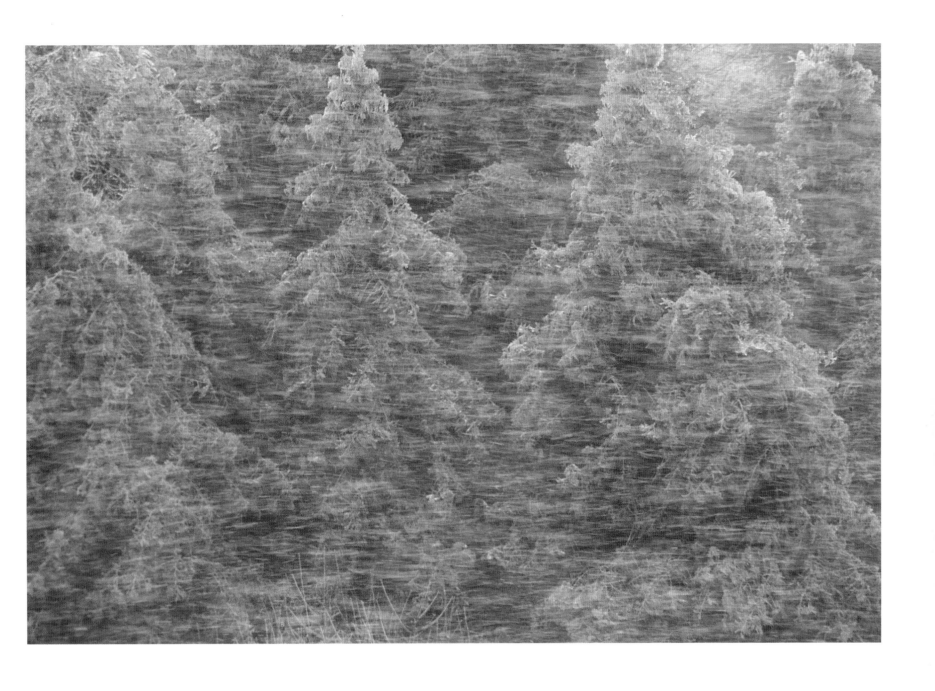

While we can only imply rapid motion and change in still photographs, we can convey it in ways that parallel or evoke our experience of it. By using a slow shutter speed at the beginning of a blizzard, before snow has accumulated on the dark evergreens, the snowflakes stand out as streamers in the wind.

Trees are social creatures; they live naturally in communities. Here they compete for light and space, and cooperate to establish conditions necessary for good health and reproduction. Left to themselves, they create and manage their environments by following natural laws, with every tree – living or dead – contributing to the success of the project.

These wilderness forests on Canada's Queen Charlotte Islands (above and opposite) are among the few remaining examples of partnership between species in Earth's virgin forests. Those stands that still exist are constantly threatened either by our ignorance of their contribution to the health of the whole planet, or by our willingness to put short-term economic gain above more important interests and values.

If we understand that we too are animals, we will recognize the plants and animals of this Vancouver Island forest as our kin, and the soil and water as the foundation of our home.

It took me a while to learn to let these alder trees speak for themselves through their lines and gentle spring hue. At first, I tried to impose a preconceived design on them by including a small evergreen tree as a centre of interest. However, the evergreen overwhelmed the delicate mesh of alders that was so compelling.

Not far from my home, there once was an old cemetery that was carpeted in spring with masses of pink and white phlox, in summer with daisies, hawkweed, and blueberries, and in autumn with goldenrod and red blueberry leaves. To me, the wild plants were beautiful in their own right and appropriate symbols of life and remembrance. The cemetery is still there, but the flowers are gone; a neatly mowed lawn is the new symbol.

We feel a certain sadness when petals fall and leaves are scattered by the wind. Yet, in its dying, every plant and animal makes room for new individuals and species that will be tested, in their turn, by their environment. Our existence is the gift of the life forms that preceded us.

Overcast conditions scatter the sun's rays, reducing contrast between light and dark tones. In an Alabama forest, the delicacy of dogwood blossoms is enhanced by this soft light and the absence of shadows. Although the white flowers may seem to be distributed at random throughout the picture space, they were in fact carefully composed to tie the vertical lines of the trees together as a visual and ecological whole.

My canoe lets me travel slowly, to explore. When I paddle the creeks and inlets of the Saint John River in early spring, I can see the land coming alive, hear the songs of birds returning from the south, smell the scent of flowers in the air, and pause as often as I want to make pictures.

In my woods in spring, tiny white violets poke through last year's leaves. The blossoms carpet the forest floor here and there and freshen the air with their fragrance. Although I go on walks simply to enjoy their brief and delicate beauty, I can never resist photographing them.

At the end of a long, hot summer day at my home, a storm may approach rapidly from the west or very occasionally from the east, as this one did. Sometimes I find the edge of the river or an inlet provides the best vantage point for viewing and photographing it.

Pictures of domesticated landscapes are portraits of our own species. Sometimes beautiful, sometimes not, they reveal our activities and behaviour, although often they tell us little about Earth itself.

This quiet, simple composition suggests a pleasant, perhaps romantic, view of farm life. Evidence of the cutting of the original forest, irrigation, and repeated use of powerful chemical fertilizers is not readily obvious, because the field appears healthy. However, the destruction of Earth's natural protective covering and the replacement of indigenous plants with a single commercial species, which is always much more susceptible to disease and insect predation, has left the soil vulnerable to damage by wind and weather.

East of Bakersfield, California, the brownish winter grasses and the rugged oak trees scattered around the hills create subdued, but memorable landscapes. Unable to find a spot where the trees and grasses formed a composition to match my impressions, I photographed each separately, and combined them later in a single image.

A morning mist softens the lines, shapes, and textures in an old bog, but it heightens the difference in tones between the darker tree skeletons and the lighter ones, strengthening the sense of depth. Completely obscuring the background, the mist evokes a sense of mystery that holds our imagination longer than if the distance were clearly apparent.

Quite often my delight in nature is a purely visual one. I am especially attracted by tonal relationships that lead the eye through and around a scene or area. For example, in the picture above, the flecks of foam scattered along the beach repeat the forms of the birds in the water. In the picture opposite, the morning sun highlights both an aspen tree and ripples on the lake.

Because of the extreme contrasts in tone, this situation required careful calculation of exposure. The tree was too small to render an accurate light reading; highlights in the foreground water caused the meter to fluctuate; so I measured the light in the dark area, then underexposed to retain its darkness. The tree's brightness and water's highlights appear here just as I saw them.

OVERLEAF   Terns flying north in a summer sky over the North Atlantic.

Attention to fine detail was the critical factor in making this pattern of wet sand resemble a male torso. By inching my camera first one way, then another, I was able to control the thickness of the black lines that define the shape. Then, by raising and lowering my tripod and zooming the lens forward and backward to find the best position, I increased the subtle variations of tone that give texture and create the impression of rippling muscles.

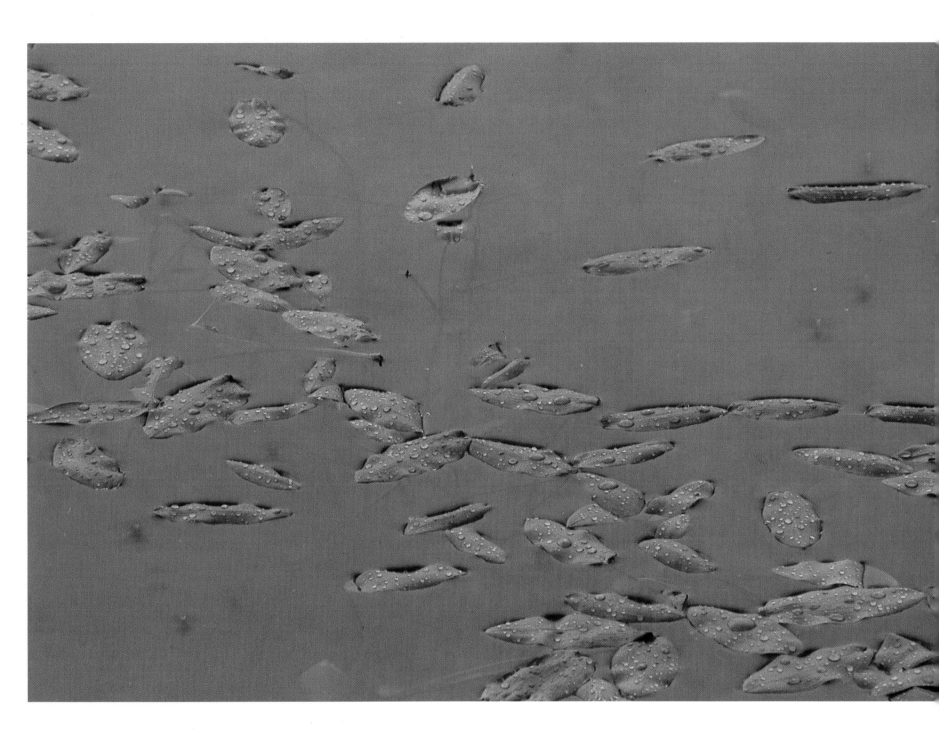

The keys to this composition are the thin black lines – the spots of surface tension where water and lily pads meet. These fragile edges separate the leaves from the continuous grey of the water, helping us pick out their faint green and brown hues. I am powerfully drawn to such subtleties and shadings in nature. I explored lakes and ponds for twenty years before discovering muted tones and hues occurring together in this way.

While driving through northern Washington state one day, I spotted this striking design in a mountain lake. The normal, placid reflection had suddenly been interrupted and reordered into a new pattern created by the wakes of swimming ducks. Caught up in a world of changing images, I continued photographing abstract designs long after the birds had flown away.

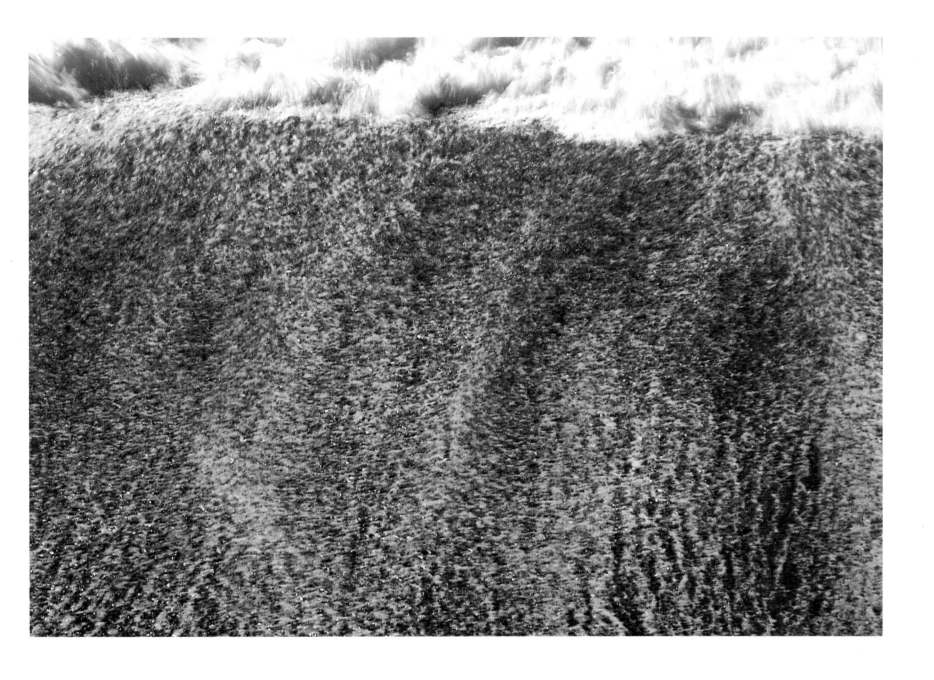

A camera, by arresting motion, gives permanence to the impermanent. This enables us to observe and appreciate the texture of a water surface at the split second of exposure. The appearance of a texture depends on the angle and distance from which we view an object or scene. From the high rock where I looked straight down on this Pacific beach, the surf had the texture of a coarse fabric as waves spread across the sand and drained back into the ocean.

Like birds, humans live *on* land or water, and *in* air. We are all sky dwellers, in differing degrees.

One evening as I stood on a high cliff overlooking the Indian Ocean, the vast expanse of the surf and the beach reminded me of how incomprehensible the vastness of time is for human beings. Looking into the distance became the visual equivalent of contemplating the past, for in the distant past, life on Earth began in the sea.

Many people find autumn a favourite time of year for walking slowly and pausing often, especially on cloudy days when the rich colours seem to glow. The patterns of leaves floating on water draw us with their contrasts of tone and hue, and with the realization that their beauty is fleeting, like that of flowers. Here, a time exposure captured the movement of leaves as they sailed around in a tiny whirlpool in a brook near my home.

On a September morning in Labrador, I wandered through the open woods behind my motel, examining the mosses, lichens, and coloured leaves of Labrador tea, and comparing them with the more familiar plant communities of my home and those of the high Arctic. Earthscapes such as these leave me with an impression of textured patterns, without any dominating lines and shapes.

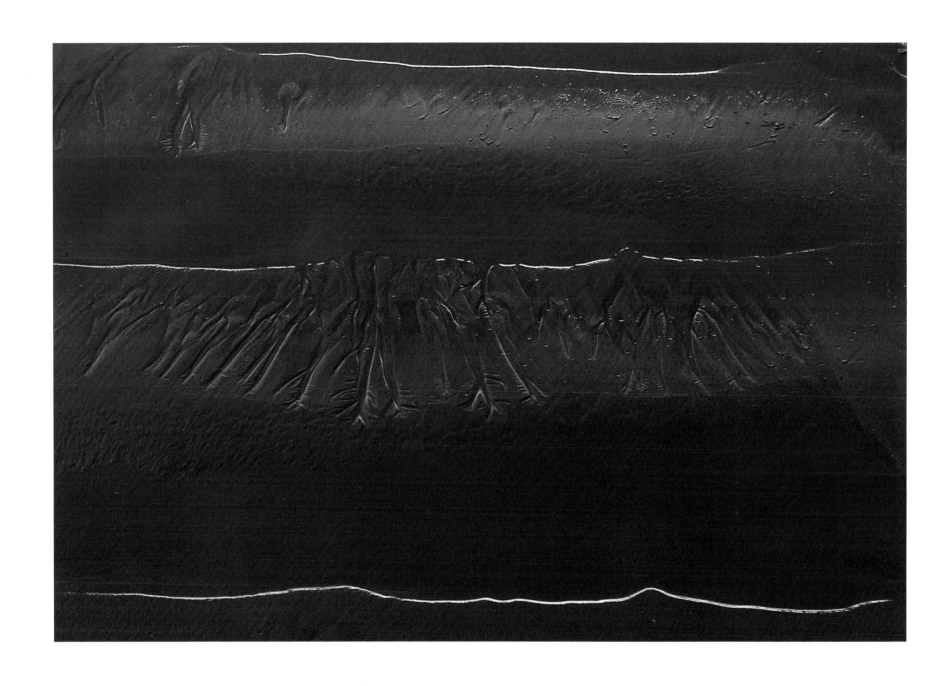

Contrasts of light create lines and shapes, and enhance texture. A puddle, a bit of beach, or a backyard is as good a place as any other to observe and photograph them.

Snow falling outside a window created a pattern of oblique lines that integrated visually with the lines and hues of tree trunks and branches. The complex arrangement produced such a simple overall impression, that I felt compelled to open the window and make the picture you see here.

With an early snowfall, two seasons seem to occur simultaneously. Because not all plants and animals are fully prepared for winter, the impact on their lives can be severe if the snow is deep and if cold temperatures persist for several days. These events create visual patterns and impressions that often capture our attention with unusual force.

When a forest is viewed from a distance, the lines and shapes of the trees and the spaces between them appear as an overall pattern. This pattern is informative as well as beautiful. It is the visual record of the competition between trees for light and space – provided by the winners.

This pattern of erosion, created by meltwater flowing down an Ellesmere Island glacier, resembles that of a very old river system. However, this deep gully developed in a relatively short span of time, as the Arctic sun shone on the glacier for several days without setting or being obscured by cloud.

The thin lines of the branches and the gentle hue of a frozen waterfall in Canada's Kouchibouguac National Park are harmonious both visually and in their emotional impact. To maintain this harmony in my photograph, I paid particular attention to proportion and tone. I established a camera position that included a greater expanse of the waterfall and reduced the size of the branches so they would retain their delicacy. Then, I overexposed to desaturate the colour.

When a major winter storm passed through New Brunswick, northern areas received heavy snow, and southern areas, rain. Midway, a band of freezing rain left trees and bushes coated with ice. Backlighting brought visual drama to ice-covered twigs and trees in the forest, while sidelighting of bushes in open areas created multicoloured, prismatic effects.

Before snow accumulates in late autumn or early winter, the floor of a mixed coniferous and hardwood forest is a complex arrangement of forms usually rendered in secondary hues. To appreciate the subtleties of nature's palette, we must stop and sit awhile, so the eye can explore the natural compositions.

On cold winter days when the river ice near my home is free of snow, I spend many hours walking or skating there, sometimes imagining that the cracks and bubbles in the ice are part of the Milky Way, and at other times simply enjoying the ice patterns without thinking of them as visual forms that have names.

PORTRAITS OF SKY AND MIND

To inquire into the intricacies of a distant landscape is to provoke thoughts about one's own interior landscape, and the familiar landscapes of memory. The land urges us to come around to an understanding of ourselves.

BARRY LOPEZ

Our Atmosphere and Beyond

In his *Pastoral Symphony*, Ludwig van Beethoven wrote what many regard as the finest musical description of "the calm before the storm," and many of the romantic composers who followed him were masters at depicting both subtle and grand variations in weather. They regarded atmospheric conditions as influencing human experience and response, and through their music deliberately evoked the emotions we feel, for example, during a raging storm or a delicate sunrise. However, composers depend on the atmosphere for more than their inspiration; they need it for the music itself. Without air there would be no music, no sound at all, because sound is caused by vibrations that compress air molecules.

Without air there would be no flowers, no birds, no people, no life at all. Air is essential to life; but not any air will do. Venus has an atmosphere saturated with sulphuric acid, a highly corrosive agent. Mars has a low surface gravity, so most of its atmosphere has escaped, and what remains becomes a violent annual windstorm that blows for months, obscuring the planet's surface. Jupiter is a giant globe of liquid hydrogen with little or no hard centre, and a surface pressure that could flatten all existing structures on Earth in the blink of an eye. However, our planet has managed to achieve what the other planets have not. It has wrapped itself in a luminous membrane of moist air that is, proportionately, no thicker than the skin of an apple. It is in this skin that all life on Earth exists. Only here do we have the right ingredients in the perfect balance to create and sustain life as we know it.

It was not always this way. There was a time before air, before life. Our planet, like all the others circling our sun, was once a huge molten mass in which the lighter substances floated to the surface, cooled, and became a solid crust. At first, there was nothing outside this crust – no atmosphere to protect it from the sun's scouring rays or the pounding of solar debris. Then, like a baby emerging from the womb, Earth gasped and began to breathe. Through volcanoes, fumaroles, geysers, and other cracks and fissures, it exhaled a mixture of gases that allowed light in and retained enough heat for life to begin.

Early life forms used substances available in the primitive atmosphere and, in turn, contributed to its development. Enormous mats of pioneer algae helped to control temperature by turning dark and absorbing heat when the planet became cool, or turning light and reflecting heat when Earth was hot. Then, a biological event of great significance occurred: microscopic blue-green algae practising photosynthesis began to enrich the atmosphere with oxygen. This started a critical period in the evolution of life. Organisms unable to live in the new atmosphere became extinct; those that survived are the progenitors of virtually all the plants and animals we know today. We live together in an atmosphere that is a gift of the blue-green algae.

The more we learn about Earth's biology, the more it appears that there is nothing accidental about the make-up of our atmosphere. Our planet is like a giant organism specifying and producing the conditions and substances necessary for life to continue, as long as the sun provides the basic energy. As part of this organism, our species is finally beginning to join with other species and abiotic elements in the preservation of this life-sustaining membrane of air.

Because air is an invisible but living environment, and because moving air brings change, in many cultures and religions wind has been regarded as the creative spirit impregnating "Mother Earth." Recently, we have learned how accurate these myths are, because wind is the medium by which many plants are pollinated, and the seeds and spores of others are carried around the planet. Wind, the carrier of genes, moves over Earth interacting with soil and water, fertilizing and nourishing all their inhabitants, accepting their gifts, and being changed in the process.

Moving air alters the topography of the land, influences the shapes of plants, and changes the patterns of our lives. In natural deserts and places where humans have stripped the land of vegetation, wind moves the topsoil, filling in depressions and building dunes. Our species generally regards this combination of erosion and deposition as being destructive. However, if we take a long-range view, we realize that nature will develop new plant and animal communities to inhabit the new landforms, as part of the amazing ability of life to adapt to change. In areas where wind buffets the land consistently from one direction, we can observe how some trees change their shape and structure to withstand the impact, and how other plants form communities in the lee of rocks, knolls, sandbanks, and even snow. While air is invisible and, therefore, impossible to photograph directly, we can show the ways air movement affects both soil and living things by focusing on snow heaped into drifts, tree branches streaming in one direction, plants crouched in rock crevices, and birds' nests hanging on one side of a tree. With a group of pictures like these we can tell the story of wind as builder and shaper.

Wind shapes us emotionally as well. The force of a gale agitates us, and prolonged, hot, dry winds can leave us apathetic and depressed. Yet, as the creative spirit, wind touches us with a sense of freedom. In the photograph on page 153, I tried to convey this spirit by showing wind blowing through a field of daisies. By using a slow-speed film and a slow shutter speed, I was able to picture the daisies tossing and dancing, perhaps like a person excited by new ideas. On other occasions, when light is fading from the sky, I use the same techniques to show rapidly-moving clouds in a symbolic way. The result may be a streaming effect that indicates the velocity and direction of the wind or, depending on my composition and the length of the exposure, a highly abstract image that invites viewers to respond emotionally to the idea of movement and change. Or, perhaps I see the shapes of clouds as the equivalent of some other physical object or situation. For example, when clouds in the daytime sky are all changing and moving rapidly at the same time, I have the sense of a landscape in motion – a collection of huge white boulders on the move. Then, I may make a sequence of slides that I can later project quickly, one after another, to evoke the rapid change.

Often I simply lie on my back in the grass studying the clouds and imagining ways of photographing them. I observe nuances of tone and hue, study the shapes, lines, and patterns, and their positions relative to land or water. Also, I try to evaluate the effect that different exposures may have on their appearance. For instance, if the bottom of a giant cumulus cloud billowing up in a clear blue sky sits on or near the horizon, I might imagine filling a vertical picture with the cloud, except for a sliver of land at the bottom. As I scan the scene, I can see that, if I move my camera position, I will be able to include a lone tree or tiny building on the horizon and thus indicate the enormous size of the cloud. Also, I may decide that slight underexposure will darken the blue sky without seriously affecting the whiteness of the cloud, thus promising a more dramatic contrast. Or, I may conclude that, because of the angle of the light, this is the ideal time to use a polarizing filter for maximum dramatic impact.

I think my favourite clouds are those associated with cold fronts. When an army of towering cumulonimbus advances relentlessly across the sky, birds take to their roosts and an eerie calm pervades the afternoon. As I wait for the hammer-blow to strike, I photograph the dark, seething mass that follows hard on the heels of the vanguard white edge – fearing, yet hoping, that I will experience the full fury of the storm. For the photograph on page 146, I used a wide-angle lens and underexposed two *f*/stops in order to retain the ominous, turbulent appearance of the approaching storm. Moments later, I was running for cover, wishing my house was not so far away.

Storms like this one, if they occur at night, provide the opportunity to photograph lightning. Before I go to a dark window or onto my deck, I always check my camera to see that the shutter-speed dial is set for time exposure and the lens is at a small opening. Then, with the camera on a tripod, I aim the lens in the direction of the storm, open the shutter, and wait for several bolts of lightning to occur. If one strikes nearby, and I feel that the dramatic lines of the single bolt may be reduced in impact by the lines of succeeding bolts, I press the shutter release, advance the film, and resume my wait. Sometimes, of course, there is no need to worry about this, because the lightning strikes everywhere except where I have aimed my lens! While my interest in lightning is mainly a visual one, every electrical storm reminds me that lightning benefits plants, and ultimately all of us, by converting atmospheric nitrogen into forms plants can use. So, I never regret those times when I fail to get a good photograph.

All clouds, not just rain clouds, are masses of water vapour condensed on minute particles of dust, salt, or ice, and like other bodies of liquid or solid water (snow, ice) have no colour of their own, but reflect the colour of the light falling on them. This is why every visually arresting sunrise or sunset depends on the presence of clouds. Perhaps no cloudscape is more spectacular than a ''mackerel'' or ''buttermilk'' sky illuminated by the glow of the sun shortly before it rises or after it sets. Altocumulus appearing as ripples form a broken sheet pattern across the sky that, in the evening, is flooded first by orange, then red, then magenta rays of light that reach a vibrant crescendo lasting for about five minutes. In the morning, the colour sequence is reversed. To capture the rich or saturated hues characteristic of many sunsets and sunrises, I underexpose slightly (up to one *f*/stop) unless the light meter suggests an exposure longer than one second for the selected lens opening. Then, reciprocity failure (the tendency of film to be less light-sensitive in weak illumination) will automatically provide the underexposure required. However, it is difficult to make a poorly-exposed picture of a sunrise or sunset. Sometimes I will deliberately overexpose a composition as much as two shutter speeds or lens openings in order to achieve a pastel effect, especially when I want to achieve harmony of impression between, say, the delicate lines of backlighted grasses and the orange or golden water behind them, or if I want to retain the lightness of a mist or haze. However, overexposing by about one *f*/stop or shutter speed is usually sufficient for situations where coloured mists form the larger part of the picture area, and less exposure will dull the lightness of the mist while enhancing the hue.

Because they rise and fall with air currents, morning mists and sea fog continually alter the appearance of an earthscape. Before sunrise on many calm mornings, I make my way through the forest behind my house to the edge of the cliff on the eastern side of the bluff. From here I can observe the first hints of colour in the sky and a panorama of distant hills, and I can look down at the long inlet below to see if mist is rising from the water surface. Often I hike down the trail to a favourite beach where I will be near, or totally engulfed by, the mist. Here I can happily spend two or three hours photographing everything from spiders' webs strung between reeds, to abstract reflection patterns, to dream-like earthscapes in light that is constantly varying its nuances of tone and hue. During my stay, the mists may change from pink to orange, to gold, to palest yellow, and finally to white as the mist parts to reveal an island, or closes to obscure hills that may have formed a background for earlier compositions, thus becoming the new background. Some mornings the mist is everywhere, and when the rising sun backlights the tiny droplets of water in the air, the river and the adjoining land are wrapped in an ethereal cloak of light. Other times I have made a series of photographs as a bank of sea fog moved steadily across deep blue water toward the land, finally enveloping me and my surroundings in a dense, white shroud that obscured everything except the nearest features of the shoreline. These are special occasions, not only because they are visually exciting, but also because they provide opportunities to photograph the important role fog plays in watering the land, especially in dry climates. Along the arid coast of southwestern Africa, where rain may not fall for years at a time, fog that rolls in from the Atlantic Ocean brings the moisture needed to sustain complex desert ecosystems. Nearly all the plants and animals have evolved body structures, or life patterns, or both, to take maximum advantage of this meagre, but essential, supply of water.

Atmospheric haze, like mist over water or wet land, scatters light rays, softening Earth's countenance. Over oceans, where haze may be intensified by spray and perhaps backlighted by the sun (as in the photograph on page 26), it can appear as an ethereal glow that evokes a feeling of mystery or appeals to our aesthetic sense with its delicate luminosity. Over land, its tendency to obscure the more distant parts of an earthscape may add to the lure of a wilderness scene by stimulating in us a desire to explore, because our imagination is held longer by things dimly seen than by those that are clearly apparent. The appearance of haziness increases with distance, so that nearer hills or mountains are darker in tone than those farther away, and serves as a clue to perspective used by photographers and painters alike. If the haze seems too pronounced, a polarizing filter may overcome the problem when the camera lens is pointed more or less at right angles to the sun, but an ultraviolet or warming filter may also be required in order to reduce the general bluishness that is characteristic of atmospheric haze. These filters can also be useful when making pictures from an airplane on a hazy day. However, I don't use them every time. To me there is a certain beauty in the milkiness that atmospheric haze adds to Earth's familiar greens, browns, and blues, and in the muting of tonal contrasts that reduces the definition of the land's structural shapes and erosion lines.

When the build-up of water molecules in the air that causes the haze becomes extreme, we may experience considerable physical discomfort and lethargy. At times like these, we long for a storm that will cleanse the air, lower the humidity, and restore our sense of well-being. Physically and emotionally, we are "air conditioned" by humidity, temperature, and air pressure. We are so wind-and-weather sensitive that we use the vocabulary of weather to describe our moods and behaviour, speaking of a "sunny disposition," a "stormy mood," and "cloudy thinking." We may "blow hot and cold" when uncertain about our feelings, or see a "ray of sunshine" during difficult circumstances, and may be "on cloud nine" when the problem is resolved. Conversely, we use the vocabulary of human activity to describe wind and weather, speaking of a breeze's "caress" or of clouds "racing" across the sky. This use of metaphor seems even more fitting when we remember that air is not a lifeless substance, but in fact is seething with life.

In addition to the other names we may give to air, it is also Earth's "protector," and not just from the sun's harmful rays. Without an atmosphere, millions of meteorites would bombard the surface of the planet every day, pounding down the hills and mountains, creating depressions large and small, and giving Earth a very different topographical appearance from the one we know. Air acts like a roof, shielding us from this steady torrent of solar debris. Only an occasional meteorite survives the friction that develops when it encounters our upper atmosphere, and streaks across the sky as a shooting star, possibly to hit Earth's surface. In order to make a safe journey through our atmosphere at extremely fast speeds, all space vehicles intended for return to Earth are fitted with special heat shields. Otherwise, they too would disappear as a streak of fire.

The most visually stunning reminders we have of the upper atmosphere are the auroral displays. The polar auroras are caused when electrified particles, shot out by the sun during periods of sunspot activity, come in contact with thin gases very high in the atmosphere. At night, I have stood for minutes or hours, often warmly bundled against the cold, to observe and photograph a spectacular display of the northern lights. Often greenish, sometimes pink or red, the auroras may form an arc or corona of colour in the nighttime sky, or leap and dance from the horizon to the peak of the heavens. Though sometimes difficult to photograph because of their movement, I have usually been able to capture them by exposing high-speed or medium-speed film for several seconds. However, since it is impossible to determine what the best exposure will be, I take the precaution of making several exposures, some up to 45 seconds in length.

Beyond the very highest auroras, Earth's atmosphere blends gradually and imperceptibly with airless space. Even in the pure blackness of a desert night, when the sky seems to be a drifting cloud of stars, the stars we see represent the merest fraction of the number we know to be there. We know as well that every star occupies not only a spot of space, but also a niche in time, and that we are seeing simultaneously different epochs in the history of the universe. How many planets have an atmosphere congenial to life? Or did have? Or will have? What sorts of life might they support? While we can speculate endlessly, we can be certain about one thing – that no planetary atmosphere can ever evolve exactly like ours, and that intelligent life, wherever it may exist, must be different from us. Perhaps more than anything else, it is our atmosphere that makes Earth's living creatures unique in the universe.

Earthscapes of the Mind

Experience or encounter with the world is central to all creative thought and activity. Creativity is not something that goes on solely within a person; it is a process or interaction between an individual and his or her world that brings something original into being – perhaps a new medicine, a new teaching method, a tool, or a photograph.

Photographers who use their own ideas and abilities to express an idea or convey a response through their photographs rather than copying the approach of somebody else, are engaged in the creative process. For example, if we encounter an earthscape that we want to photograph, we will observe it from many positions, study the way lighting affects the natural designs, and become absorbed in it. Sometimes we will decide to alter the form of the subject matter very little and concentrate on selecting and emphasizing (by our own particular combinations of design and photographic techniques) the visual elements that stimulate our emotional response. Even though the final image may closely resemble the original material, trying to evoke the same response through a photograph may involve a great deal of creative effort (see the photograph on page 17). At other times we may see the subject matter as the "equivalent" of something else, and choose a camera position and photographic techniques that enhance the resemblance so that viewers will clearly see and respond to the material in the same way. For instance, in the photograph of sand on page 116 the main shape is that of a male torso, with the ripple patterns resembling muscles. Even though we see an image of sand, our main response is to its likeness to the human body rather than to the actual subject matter.

On still other occasions, an earthscape may stimulate pictures in our mind that are quite unlike the original scene, but which form an important part of our response to it. While these *earthscapes of the mind* are based on our experience of the physical world, they have no actual counterpart in that world. However, by using our knowledge of design and our technical skills to photograph them, we bring into being images for everybody to see. For example, in the photograph on page 145, I have combined (by double exposure) an image of a cedar tree with another of lupin blossoms. This tree is a favourite of mine. It stands in the field below my front deck, and I often make photographs of it from that viewpoint. In June, the meadow in front of the tree is filled with lupins in full bloom. When I sit on my deck I enjoy the beauty of both the tree and the flowers. Although quite separate from each other, they appear simultaneously in my field of vision and form a unitary impression; they become a single, forceful visual experience. So, I decided to unite them in a single image – a new earthscape – in order to give others a sense of what I experience.

It is helpful to think of earthscapes of the mind in terms of dream imagery. Dreams are part of our visual imagination. When we dream we often see the physical world in ways that give us new perspectives on the actual world of our waking lives. In our dreams we rearrange the parts of an earthscape freely. A familiar house sits on an unfamiliar hill; we struggle out of a dark, frightening jungle onto a broad, sunlit plain; we see a patch of colourful flowers blooming in the snow. Similarly, we observe Earth from positions that are inaccessible to us in real life. We fly around at will, observing situations from various heights; we see through a wall as clearly as we see through windows.

In our conscious dreams or reveries we also conceive of imaginary earthscapes. For example, before we visit a place for the first time, we picture it in our mind (often in clear and vivid images that may be very unlike the reality), and we accept this previsualization as a perfectly normal, indeed necessary, kind of mental activity. However, unlike the dreams of our sleep, reason constrains our imagination, and we become concerned with limits or form, because we have to operate in the context of the real world where freezing weather prevents plants from blooming and where communication with others is necessary and desirable. So, instead of dreaming of gardens in the snow, we imagine transforming an abandoned quarry with vines and flowers or some other action that, in theory, is possible for us to complete, even if the time and resources are not immediately available.

The visual forms or limits that reason imposes in conscious dreams are similar to the photographic experience. We create our pictures by focusing our lenses on already-existing objects and their forms. No matter how much we may restructure, distort, combine, or otherwise alter the appearance of the things we photograph, we must consider their visual forms and work with them – just as a gardener may have to work with the shape of a quarry and the colours, lines, and shapes of plants. While it is necessary, often exciting, and sometimes just plain good fun for photographers to experiment with various techniques, the images that result from experimentation alone often fail to convey any real feeling or meaning. Genuine earthscapes of the mind, like all genuine art, reveal a caring encounter with the subject matter. Such images are tangible visual expressions of a significant response that viewers can share when they see the photographs, provided they are willing to engage in the creative process themselves, to be open to new or unusual visual constructions.

The creative process requires photographers to use both their imagination and their analytical ability. The pictures we imagine must be analysed in order for us to take the sequence of steps that will transfer them to film. For example, for the picture on page 147, I wanted to make a personal statement that had emerged from several experiences I had had in which *good ideas were so rigidly held that they became destructive*. For me these ideas were masked monsters, like creatures or flowers that lure prey with their beauty only to devour it when it comes too close. I wanted to create a visual form that was both beautifully symmetrical and disturbing – one that had the potential to attract and repel at the same time.

My next step was to search my immediate home environment for something that would express my mental image. It had to be beautiful, with a highly-structured and regular form. I didn't *know* exactly what I wanted, but I could *feel* it very clearly indeed. One day when I was weeding my vegetable garden the form for my picture leapt out at me. I sat down beside the row of ornamental cabbage plants and studied them carefully. I soon realized that even the most elegant one could serve only as a *basis* for my photograph. Now I had to consider techniques that would transform the symmetrical beauty of the cabbage into a new form that would retain its attractiveness yet evoke a mysterious feeling of threat.

I began by setting up my equipment and experimenting with composition and exposure. My first close-up compositions showed the natural beauty of the cabbage leaves, but failed to function as equivalents for my idea because they were insufficiently abstract. Before long I realized that, while retaining the beauty of the leaves, I had to eliminate completely the label or idea of "cabbage." Also, I needed to underexpose in order to create an ominous sense of darkness. After some more trial-and-error, I decided to make two identical, slightly overexposed and partly out-of-focus images of the cabbage, and then to sandwich them together, reversing one in the process. This would produce a slightly under-exposed, highly symmetrical, Rorschach-type form that was both elegantly attractive and mysteriously threatening. Thus, the image that began as a vaguely-conceived picture in my mind was formed by using existing shapes, lines, textures, and hues in new ways to create the feeling I wanted. Because everybody is unique and has unique experiences, no two photographers will ever imagine exactly the same pictures or use the same design and technical skills in exactly the same way. So, each person can produce photographs that are genuinely new and special creations.

However, the creative process of making photographs (whether of mental images or physical things) does not begin in the imagination – but in observing the world around us. We have to see a suit before we can imagine owning one; we have to observe the stars and planets in a midnight sky before we can imagine bubbles in ice or glass as their visual equivalents. Observation is the key that unlocks our visual imagination.

Observing things *without naming or labelling* them is fundamental to all visual art. Labelling, which is seeing with words, interferes with pure observation – with seeing things as they are. Often when I am writing I consciously observe my coffee mug without thinking of it as a "cup" or "mug." Immediately, I find myself wandering effortlessly through its wonderful surface patterns and textures – enjoying the scenery, as it were. My mug was fired in a wood-burning kiln, and its glaze is a rich, earthy brown smattered here and there with rough, ashy freckles and tiny, creamy-blue swirls. Roaming visually over the surface, I make no attempt to label the imperfections or other variations in the glaze as "depressions," "streams," "constellations," or anything else, but simply observe them as colours and tones. In the process I am providing a storehouse of visual memories for my imagination to draw on when I am dreaming or making pictures. Moments later, when I think in labels, I may note that the patterns on the coffee mug resemble the Namib Desert or the surface of the moon. My memory then files my experience under several headings – "cup," "desert," "moon," "texture," "visual equivalent" – for easy access when I need or want them. Similarly, when I am standing beside a large clump of tangled ferns in the field behind my house, I may deliberately forget "fern," "tangle," and other labels for a while so that I can enjoy a purely visual experience. The act of seeing without words takes conscious effort at first, like learning to ride a bicycle. However, with a little practise, it becomes so effortless that we can make the transition from one mode of seeing to the other more easily than hopping on or off a bicycle. Pure seeing means that both our more realistic pictures of Earth and the earthscapes of our mind begin where they should – with clear vision. If we begin well, then when we engage our imagination and our intellect in the step-by-step process of composing, choosing depth of field, and exposing, our final image is much more likely to communicate the information or to express the feeling that we want to share with others.

Although both the photographs we make of actual earthscapes and those that we create of imagined places begin with observation of our world, for imagined earthscapes we add a step between the observing and the picture-making. That step is sometimes delayed for a very long time, and may be clearly influenced by many experiences and observations. When we carefully examine a scene, we almost inevitably store it in our memory. Months and even years later, it may surface in our consciousness as representing a mood or an idea, perhaps because of its dark, ominous tones, its saturated hues, or its unusual shapes. If we have developed a storehouse of observations, other similar scenes will also come to mind, and from these our imagination can fashion a new earthscape that clearly expresses our thoughts and feelings.

Making photographs of imagined earthscapes is as important as conceiving them, because then we exercise our imagination still more and gain greater control over technique. Imaginary earthscapes extend experience. They are no less real than physical objects or other ideas or conceptions, and by making them into photographs we can communicate with others just as surely as if we were using words, and often more effectively. Even when we fail to produce the pictures we want, as happens from time to time, we will almost invariably discover that the experience of trying produces new ideas or directions that are worth pursuing. In fact, sometimes our failures can be more exciting and more useful in the long term than our successes. It's as if creative effort always brings a reward, though not necessarily the one we hope for or expect.

On my regular walks through my woods and fields I always feel that, even if I do not return with better pictures than I have made before, I am a better photographer for having been with the trees, the bogs, and the creeks that I thought I had already observed and come to know so well. On every walk they surprise me once again with some aspect of their individual natural design or community structure, adding to my knowledge of natural relationships and making me more aware of the relationships between myself and members of my own and other species.

In June, the candle-like blossoms of lupins cover a field between my house and a large cedar tree. A double exposure enabled me to convey my impression of the scene in a single image.

We live at the bottom of an ocean of air. It is here, in the lowest level of the atmosphere, that all weather occurs. Storms and winds distribute heat and moisture around the planet, making most of Earth's surface hospitable to life.

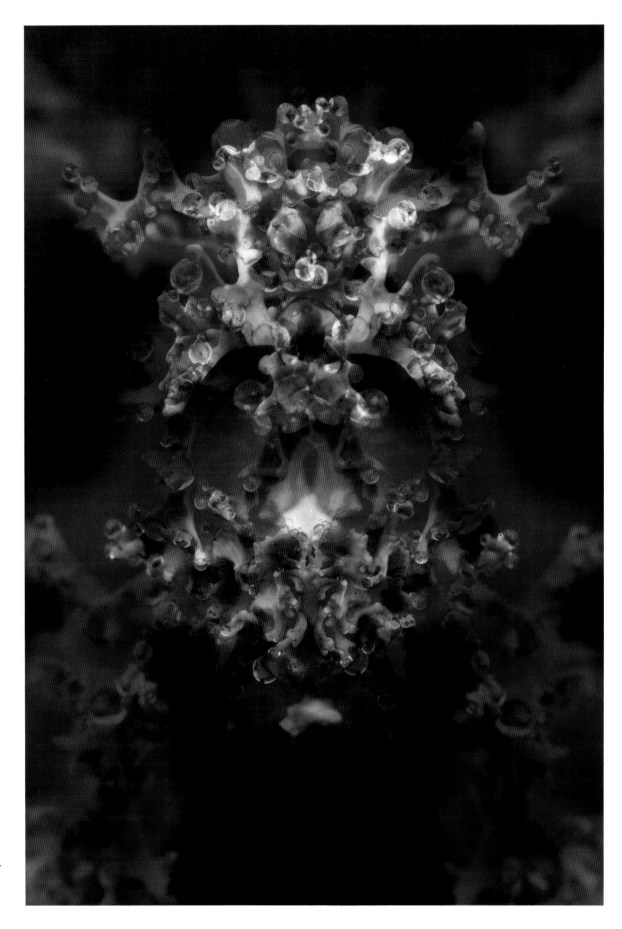

Every year, I plant a row of ornamental cabbages. Unlike the other cabbages in my garden, these are not intended for my table. Instead, I spend hours each season photographing the patterns in their leaves, as in this picture, treating the cabbages symbolically.

This double exposure is not intended as a literal document of a summer meadow, but rather as an expression of my feelings of delight and exuberance when I was there.

For a few days every spring, flowers carpet this Namaqualand valley on the southern fringe of the Namib Desert; and for only a few minutes on each of those days, the afternoon sun lights up these bands of flowers with a special brilliance, then disappears behind a mountain.

OVERLEAF Spectacular displays of wildflowers are an annual event in Namaqualand. Because the right conditions for germination, growth, and reproduction occur mainly in August and September, hundreds of annual species and most perennials flower simultaneously.

Summer in eastern Canada and the northeastern United States is usually characterized by lush vegetation and flowers blooming from early April to late October. Along roadsides, wild plants are often mixed with species that have escaped from gardens, such as these orange daylilies.

Most plants are accustomed to wind, and spend much of their lives tossing, dancing, and bending. Wind pollinates many species, sometimes transporting the seeds and spores over great distances.

For me, nothing surpasses the experience of being on a wilderness lake or river at sunrise. Here, away from the hustle and bustle of human activity, we can observe less hurried societies beginning the day and hear the natural sounds of Earth, an experience that makes us reassess the patterns of our own lives.

Careful observation of nature must engage all our senses, so we not only see the reflections of sky and trees on the surface of a pond, but also feel the gentle breeze on our face, smell the mustiness of old leaves decaying on the forest floor, hear the songs of warblers, and taste the delicate sweetness of a wild strawberry.

If we could enter a time capsule and go back a million years or so, we would probably be able to make a photograph very similar to this one, of life establishing itself on a rock face – a natural, visual representation of the dynamic, expanding universe.

If we can remember that we are only one of millions of species that inhabit this planet, and that we are part of an intricate network of interrelationships between all natural things, then perhaps we can recognize the importance of each of the other species to the balance and existence of the whole, and the necessity of preserving all parts of a natural habitat, such as this one.

I have climbed this lichen-covered koppie (rocky hill) in Namaqualand many times, and no matter how early in the day I commence my hike, only the darkness of night brings me back down. The extrusion provides no pasture or garden land for the farmer who owns the hill, but for photographers and lovers of wild places, it is a paradise. Hundreds of plant species are found here, as well as insects, birds, reptiles, and small mammals. The visual and ecological complexity of the koppie is stunning.

On another farm not far from the rocky hill (opposite), periodic desert rains tumbling over a cliff for thousands of years have polished the Precambrian granite with granules of sand. The rock has become a mirror, reflecting the sun's silver rays at midday and glowing orange in the sunset. Like the koppie, its continued existence in this wild state is the happy result of benign neglect.

For only a few days a year, and for only a few moments on those days, the last pink rays of the sun reach into this remote corner of an Orange River canyon. So brief was the gentle illumination on this day that I had time for only two exposures before the light faded.

On Nova Scotia's Brier Island, at the mouth of the Bay of Fundy, huge basalt columns are the geological remnants of an ancient volcano. High on the columns, collections of wildflowers hang from crevices, while in the intertidal zone at the bottom, the rocks are studded with barnacles and sea algae. Late on sunny afternoons, when the columns are in the shade, light reflected from the water and sky gives the basalt a bluish tint.

When we make realistic images or documentary photographs, we should try to let the subject speak for itself through its natural designs, and avoid imposing preconceived designs on our compositions.

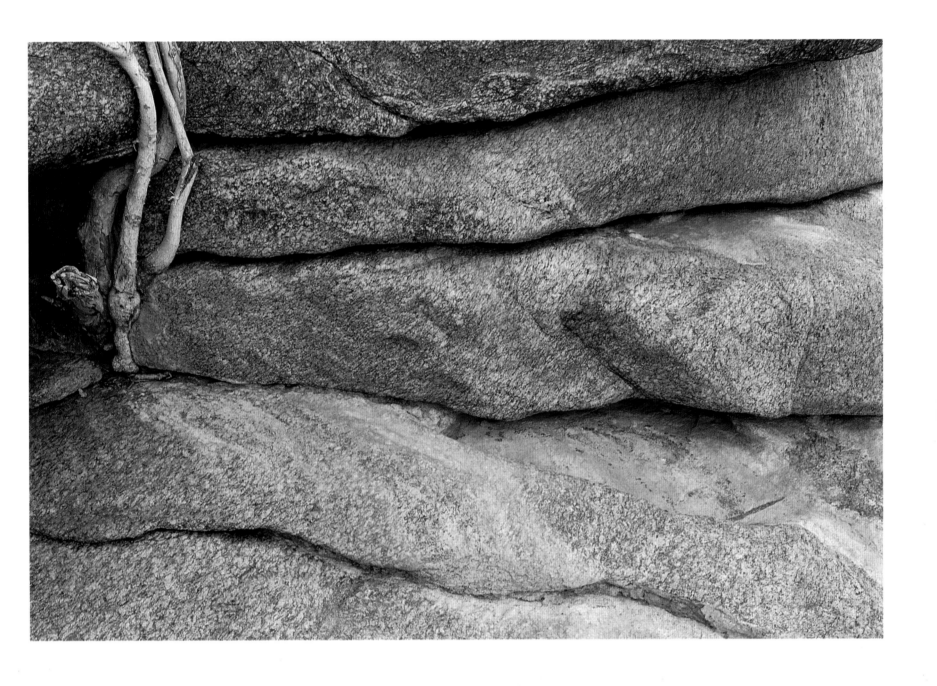

For the photographs above and opposite, I spent a long time studying the natural designs to gain information about the habitats and to determine why these situations had so attracted me visually. In making both designs, I decided that my primary task was to select those elements that would show the situations simply and clearly, enabling viewers to decipher the natural history and respond to it in their own way.

The play of light and shadow creates graphic forms that obscure, overwhelm, or emphasize Earth's natural designs, concentrating or redirecting our attention in the process. Here, the large black shadows made me focus my attention on the triangle of sunlit rock. In turn, this shape led my eye directly to the water, making me far more aware of the stream than I had been earlier in the day when the sky was overcast.

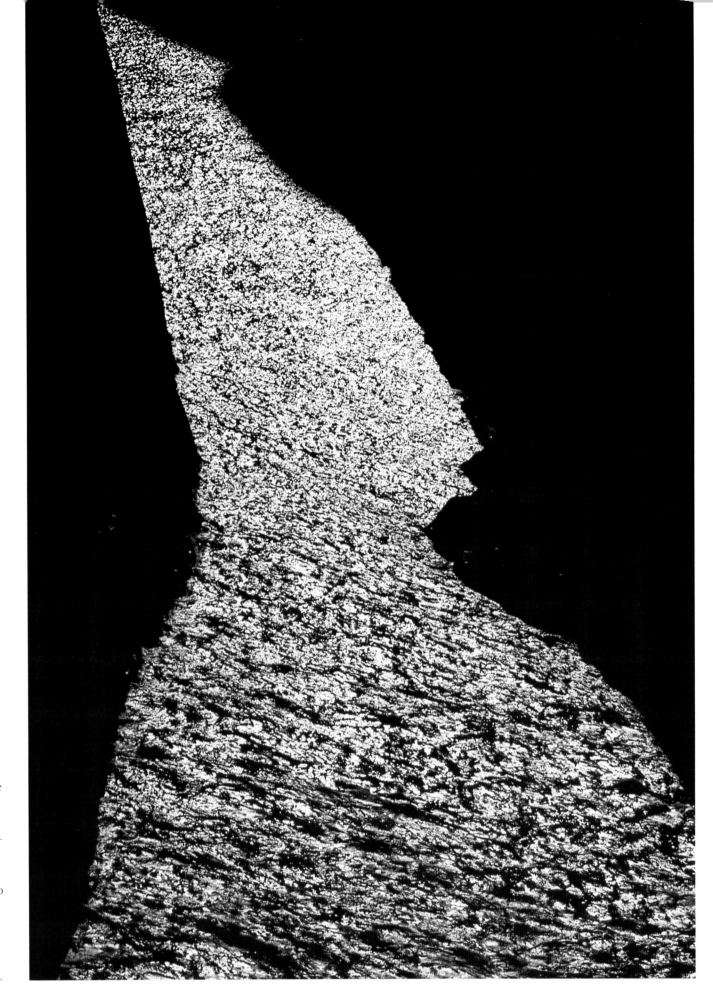

Both of these pictures are about water. In the picture opposite, I have used the central triangle of rock to lead to the flowing water – the final point of interest. In this photograph, while the central form also leads us through the picture space, it does not lead to anything beyond itself – because the water is both content and form.

A road is a line that leads the eye. However, when the line is obscured before it reaches a clear destination, it arouses our curiosity and appeals to the imagination.

Situations like these are given to us only for a moment. We can see and photograph them if we prepare ourselves by regularly observing the moving configurations of light and shadow.

In the high Arctic, mountain ridges tower above a vast expanse of snow and ice. Life seems impossible in environments like this. Yet, jumping spiders have been observed hopping across the ice high on Mount Everest. They are fed by the wind, which airlifts in an appetizing selection of insects. In other extremely cold places, small forms of animal life dine on pollen, spores, seeds, and plant fragments, also delivered by the wind.

The shapes of these steep hills in northern Washington state have been determined by both tectonic forces and erosion. Although wind and water will inevitably carry away some soil, grasses control the speed of erosion by anchoring the soil with their roots and by depositing organic material from last year's grasses that contribute to the creation of new topsoil. Grazing by sheep and goats destroys the grass cover and results in the rapid loss of topsoil and development of deep gullies.

Over the years, hundreds of thousands of people have climbed to this vantage point for a view of Moraine Lake in Canada's Banff National Park, and millions of others have seen photographs made here. Yet, each photograph is slightly different. Sometimes the lake is as smooth as glass; at other times wind ripples all or parts of the reflections or makes the water so rough that reflections disappear entirely. And, each year, this tree grows a little taller. Visually, we can never return to exactly the same spot.

Perhaps even more people have stopped by Bow Lake in Banff National Park to view the reflections of the snowbanks, the jumbled rocky scree, and the trees growing at its base. Often, the air is so clear that distance is deceiving. Only the trees give a clue to the size of the lake. However, in vistas like this, scale seems unimportant. Together, the land and the reflections form a striking, symmetrical, abstract design that is visually exciting in its own right.

One overcast day I sat beside a waterfall in the woods near my home watching the water tumbling and splashing. After a while, my gaze shifted to the pool at the base of the cascade, where I searched for submerged rock patterns. Gradually I became aware of the pattern in the moving surface. This composition contains no lines or shapes that would distract attention from the water pattern, the real centre of interest.

My approach to this composition was the opposite of that for the picture on the facing page. Although the cloud pattern first attracted my attention, once I spotted the black tree skeleton I decided to include it as a centre of interest, because it emphasized the forms of the clouds and the direction in which they were moving. At the same time, the tree suggested the inevitable result of prolonged drought, arousing a feeling of uneasiness.

Not long after a rain carries soil into an arid or desert area, the film of mud deposited on the bottom of temporary streams and puddles begins to dry out and contract. Hard mud tiles form patterns, which may last for years. When the wind blows, sand is scattered over the tiles and into the crevices between them. This beautiful natural mosaic reveals the entire process.

Anywhere that snow falls or frost and ice form, it is possible to make voyages of pure fantasy, or to explore worlds that can be seen only with the aid of a camera. For this picture, I used a macro lens set at minimum focusing distance and widest aperture, and inched my way along the edge of a melting Arctic snowbank. I have made similar photographs at home on an icy window ledge and in snow melting on the roof of my car.

Any configuration of a wave can only be photographed once. We never have a second chance. However, arresting the constantly changing designs of waves to capture the best compositions is the result of careful preparation, not luck. During a week in the Bahamas, I spent several hours every day observing and photographing waves breaking over a sandy beach. Gradually, I learned to visualize the expressive peak of a design a moment or two before it occurred, and to press the shutter release with split-second accuracy. On the final morning, I made this image.

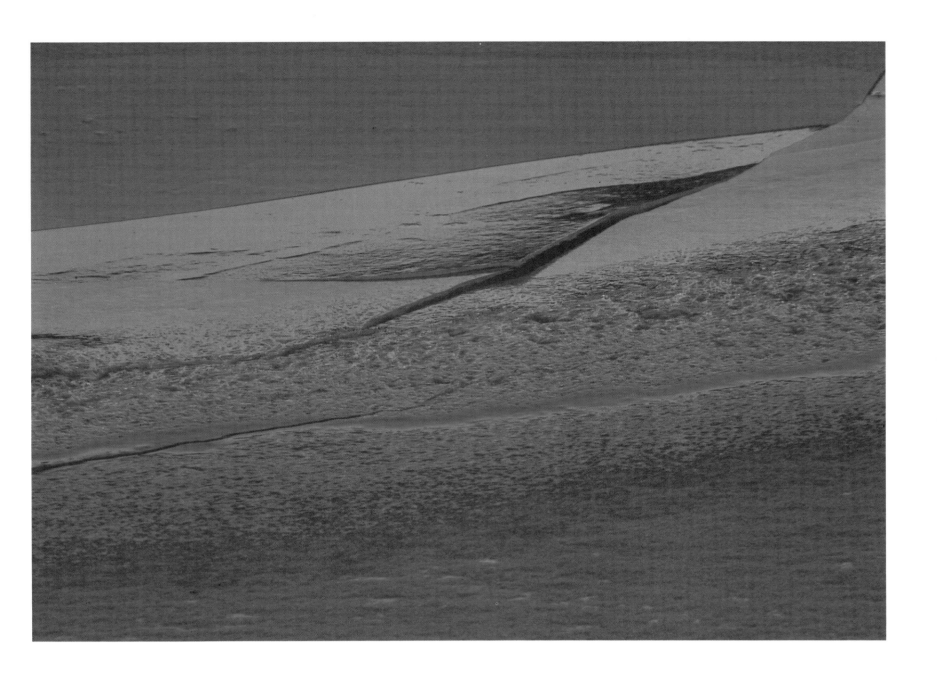

Rising long before dawn on a winter morning, I bundled up warmly and headed for a frozen river near my home. As the sky turned orange and peach, I made overall pictures of the river curving away to the horizon. Then, I began to focus on patterns of ice. Peering at the new earthscapes I had discovered, I began to reflect on how quickly they would disappear as the temperature rose or after the approaching group of snowmobiles passed by.

In the long history of Earth, there have been only a few times when a particular event or configuration of events has had the capacity to alter radically and forever the future of our entire planet. Yet, it is precisely at such a juncture that we now stand, and know that we stand. Unlike any species before us, we have the knowledge and the means to determine the future of Earth. But do we have the will to choose wisely?

We have begun to contemplate our origins: starstuff pondering the stars . . .
tracing the long journey by which, here at least, consciousness arose. Our loyalties
are to the species and the planet. We speak for Earth. Our obligation to survive
is owed not just to ourselves but also to that Cosmos, ancient and vast, from
which we spring.   CARL SAGAN

ABOUT THE PHOTOGRAPHS

The photographs in this book focus on Earth's varied topography and environments. While they were made in locations as widely separated as the Arctic's Ellesmere Island, the Caribbean's Eleuthera Island, and southern Africa's Namib Desert, the pictures on pages 13, 36, 97, 103, 126, 130, 131, 145, 146, 148, and 153 were all made within a few steps from my home, on the Saint John River, in New Brunswick, Canada. Several other photographs were made only a short walk or drive from there. By regularly observing the various natural environments around my home, they have become like old friends to me.

When I am working at home, I often seek out and photograph small areas that are quite unlike the general environment, but which resemble places I would like to visit, or have already visited and would like to visit again. For example, by making repeated trips to a local gravel pit over the years, I became familiar with desert habitats and designs, and prepared myself for some of the visual challenges and opportunities presented by major desert ecosystems. Conversely, several visits to distant environments, such as the rain forests of the Pacific coast, helped me to observe and understand my home forest in new ways.

Wherever I travel I carry a sturdy, medium-weight tripod, which I use constantly to ensure careful composition. Except for two photographs made from my canoe and three others of birds in flight, all the pictures in this book were made with a Minolta 35mm camera and lens mounted on my tripod. The lenses I used most frequently were a 28-85mm and a 70-210mm zoom; the films, either Kodachrome 25 or Fujichrome 100. Because light conditions are constantly changing, I never record exposure data, but concentrate instead on observing the different areas of tone in the picture space, reading the light meter in my camera accurately, and then deciding if any overexposure or underexposure is required. Except for two double exposures and two "sandwiches," which are indicated either in the captions or the text, none of the photographs has been mechanically or chemically manipulated. However, I often used a warming filter to counteract the bluishness of ultraviolet light and, in the Namib Desert and for long-distance photographs on Ellesmere Island, I employed a polarizing filter to eliminate the negative effects of spectral light.

Photographic equipment and techniques are to a picture what a car or airplane is to a trip – they get us there, but they are not the reason for our journey. For me, the importance and the satisfaction of photography lies in carefully observing the subject matter and how it is affected by light, then in selecting and composing aspects of it for presentation as an expressive image.